PRAISE FOR **HENRY PETROSKI**'s

REMAKING THE WORLD

"Highly readable. . . . What Petroski does best is to make civil engineering understandable to a general audience. . . . This book is guaranteed to make any reader appreciate the work these creative professionals do for us all." —*Rocky Mountain News*

"One lays this book down with a renewed appreciation for the work of all these 'men and women of progress,' and for the grace and insight of the writer who tells their stories so well."
—*The Sciences*

"[*Remaking the World*] is a great introduction to the world of engineering and an education that ensures you will never regard the house in which you live, the roads upon which you travel or the technology with which you work in the same ways again."
—*Post & Courier* (South Carolina)

"A rewarding read . . . the book's charm lies in the countless anecdotes and bits of historic and engineering trivia that pepper each essay, rich details guaranteed to stay with you." —*Forbes*

HENRY PETROSKI

REMAKING THE WORLD

Henry Petroski is the Aleksandar S. Vesić Professor
of Civil Engineering and Professor of History at Duke
University, where he also serves as chairman of the
Department of Civil and Environmental Engineering.
He is the author of seven previous books.

HENRY PETROSKI

REMAKING
THE WORLD

ADVENTURES IN ENGINEERING

VINTAGE BOOKS

A DIVISION OF RANDOM HOUSE, INC.   NEW YORK

FIRST VINTAGE BOOKS EDITION, JANUARY 1999

The Library of Congress has cataloged
the Knopf edition as follows:

Petroski, Henry.
Remaking the world: adventures in engineering / by Henry Petroski.
p. cm.
Includes bibliographical references and index.
ISBN 0-375-40041-9 (hardcover: alk. paper)
I. Engineering. I. Title.
TA145.P47 1997
620–dc21   97-29328   CIP

Vintage ISBN: 0-375-70024-2

Author photograph © Catherine Petroski

www.randomhouse.com

Printed in the United States of America
10 9 8 7 6 5 4 3 2 1

TO MY SISTER, MARIANNE

*The scientist seeks
to understand what is;
the engineer seeks
to create what never was.*
—attributed to
Theodore von Kármán

Contents

Preface

In 1986 Sigma Xi, The Scientific Research Society celebrated the centennial of its founding as an honorary society for scientists and engineers, somewhat akin to Phi Beta Kappa. Today the organization of some ninety thousand members fosters interaction among science, technology, and society, encourages the appreciation and support of original work in science and technology, and honors research accomplishments. As part of the centennial, the Board of Directors of Sigma Xi began to look to its second century, and it was decided that a relocation of its headquarters from New Haven, Connecticut, was in the best interests of the society.

The move was completed early in 1990, and it brought the editorial offices of *American Scientist*, Sigma Xi's bimonthly magazine, to Research Triangle Park, North Carolina, where its new editor, Brian Hayes, assembled a new editorial staff. At that time Brian wrote to me that he was looking to broaden the coverage of engineering and technology in the magazine, and he asked if I might send him some material for consideration. I did send him a couple of essays that I had been working on, and he then approached me with the proposition of writing a regular engineering column. I agreed to do so, and this book is by and large a selection from the three dozen or so columns I have now written for *American Scientist*, along with three essays first published elsewhere, and in some cases with material restored that originally had to be cut because of length restrictions in the magazine.

I am grateful to Brian Hayes for giving me the opportunity to write for *American Scientist*. I am also grateful to Rosalind Reid, Brian's successor as editor of the magazine, to Mike May, and to David Schoonmaker for their helpful and sensitive editing of my manuscripts and for locating and selecting the illustrations that have accompanied almost all of them. Indeed, it has been a pleasure to work with the entire staff of *American Scientist* in the course of seeing each manuscript through the press. It has also been a pleasure to work once again with Ashbel Green, who helped select the essays included here, and with Jennifer Bernstein, Melvin Rosenthal, and others at Alfred A. Knopf involved with editing, designing, and producing this book. And, as always, I am thankful to the many librarians at Duke University and elsewhere who secured essential materials for me, sometimes under the urgency of deadlines that come with a regular publishing schedule.

Among the joys of publishing in a magazine that is read by a widely diverse group of people is the feedback that comes as each issue appears. I am grateful to my many colleagues at Duke and elsewhere who have commented on the columns and who have encouraged me to continue to write them. There has also been a steady stream of mail from new and increasingly familiar readers that expands on and often adds their own personal experiences to what I have written. I am grateful for the many helpful and interesting letters from these readers, but there have been too many of them for me to acknowledge individually here. Where I have incorporated their information into these revised and amplified versions of the columns, I have expressed my gratitude for their contributions in the Bibliography and Acknowledgments section at the back of this book. In some cases I have also elaborated there on aspects of a topic that have seemed too tangential to insert into the essay proper. In all cases I have noted the date of first appearance of an essay and have listed my principal sources for quotations and information.

Finally, and as always, I am grateful to my family. My son, Stephen Petroski, helped me by extracting information from the files of my old columns. My daughter, Karen Petroski, has been an excellent sounding board. And my wife, Catherine Petroski, has been a faithful first reader of all of these columns, even as she has been engaged in her own research and writing.

Durham, North Carolina
May 1997

REMAKING
THE WORLD

IMAGES OF AN ENGINEER

When I was an engineering student, a friend of a friend nicknamed me "Steinmetz" and refused to call me by any other name. This person, whose inclinations I recall to have been toward business and politics and not at all toward science and technology, had learned from his parents that Charles Steinmetz was the embodiment of American engineering. Until I became his namesake, however, I had not heard of Steinmetz. At first I bristled at being called by a name I did not recognize, but when I was told that Steinmetz had been "a famous engineer" who gave "great lectures" that had enraptured audiences, I was flattered and enjoyed the sobriquet. That friend and I went our separate ways, and many more years passed before I learned more about Steinmetz and understood why, to so many Americans—even those who knew little about technology—his name came to be synonymous with engineering.

Karl August Rudolf Steinmetz was born in Breslau, Germany (now Wroclaw, Poland), in 1865 and excelled in classical literature, physics, and mathematics at gymnasium. At the University of Breslau he completed all the course work and a thesis for a doctorate in mathematics, but his extracurricular activities among the young Socialists during Bismarck's crackdown forced him to flee to Zurich, Switzerland, before receiving his degree. In 1889 he immigrated to America, where he simplified his given names to Charles Proteus, the latter after a school nickname that presumably referred to the Greek sea god who could change his appearance at will. Steinmetz lived in Brooklyn and commuted some distance to Yonkers to work for Rudolph Eichemeyer, an exiled German Socialist turned capitalist and a manufacturer of electrical equipment. Steinmetz began working for $12 a week, and he drew on his mathematical physics background to solve practical problems in electrical engineering for the firm, which was soon bought out for its patents by the General Electric Company. As an adjunct to his employment, Steinmetz joined the American Institute of Electrical Engineers and in the early 1890s presented before that society papers that showed him to be in the forefront of the field. His law of hysteresis loss, for example, enabled engineers to calculate—and so minimize without costly trial-and-error experiments—the power losses in transformers that were hindering the distribution and use of electricity.

Steinmetz began his long association with General Electric in Schenectady, New York, in 1892. He was first put in charge of the calculating department but soon rose to chief consulting engineer, a position created especially for him. According to George Wise, who has written extensively on the history of industrial research at General Electric, in this role Steinmetz was to troubleshoot major technical problems that arose at GE and to advise top management on technical issues and opportunities. Toward the close of the decade he began advocating the establishment of a research laboratory in Schenectady, and in 1900 Steinmetz finally convinced the company to start up an electrochemical research laboratory that would develop new types of lamps and thus help GE deal with the growing competition, especially from more engineering- and science-based companies such as Westinghouse.

The position of consulting engineer left Steinmetz a good deal of time to lecture, teach, and write. The proximity of Union College gave him the opportunity to lecture on the theoretical elements of electrical engineering, and it was from Union that he was to be awarded, albeit *honoris causa*, the doctoral degree that he did not get in Germany. Steinmetz had developed and introduced the use of complex numbers for calculations involving alternating-current circuits, and his textbooks on such techniques and the mathematics to effect them provided the theoretical foundations for generations of engineers. His pioneering engineering work gave Steinmetz free rein to work according to his own style at General Electric, and his eccentricities became as famous as his achievements. His collection of grotesquely shaped cactus plants, an ever-present cigar that mocked GE's "No Smoking" signs (some said to have been changed to read, "No Smoking. Except Dr. Steinmetz"), and the bathing trunks and red sweater in which he received visiting dignitaries at the retreat where he often preferred to work became symbols of his genius.

Steinmetz's retreat was an isolated camp on a tributary of the Mohawk River. He had the small stream dammed to form a placid pond, and more than one writer commented upon Steinmetz's use of it. A journalist who visited wrote, "Probably there is no other office like it in all the world—a battered twelve-foot tippy canoe with a cushion in the bottom and four boards laid together from gunwale to gunwale, thwartwise to serve as a desk. When he goes down to the river to work he carries his papers under his arm, with them Hutchinson's volume of four-place tables, and a little Nabisco box wherein he keeps his pencils." The tables of logarithms often remained unopened, and Steinmetz was said to have taken them along only in case his memory failed him.

By the turn of the century his fame was almost without bounds. In 1902 Harvard University honored Steinmetz with a Master of Arts degree, and he was reported to have attended the commencement ceremony in "ordinary business clothes, which most men would have been ashamed to go fishing in." But he was hardly being honored for sartorial excellence; Harvard's president, Charles William Eliot, conferred the degree upon Steinmetz as "the foremost electrical engineer in the United States, and therefore the world."

Steinmetz's achievements and eccentricities made him a natural subject for stories in newspapers and popular magazines, which at the time took science and technology as seriously as they

Steinmetz working in his canoe

did other aspects of news and culture. According to Marcel LaFollette's *Making Science Our Own: Public Images of Science, 1910–1955,* Steinmetz ranked among the most frequently appearing subjects or contributors of stories about science and technology published in popular American magazines of the period. It is not without design that Steinmetz and Thomas Edison, who tops LaFollette's list, achieved such public prominence, for their association with electricity, and with General Electric in particular, was exploited by the corporation to the utmost. The exploitation, however, may have had an unintended and unfortunate side effect on the shaping of the popular image of engineering in America.

The public-relations department of General Electric seized upon the popularity of Steinmetz's image and promoted the engineer who first generated the power of lightning in the laboratory as "the very personification of modern electrical development," thus giving the cold, abstract, and sometimes feared new technology a human embodiment. Steinmetz's fame, eccentric lifestyle, and cooperative nature provided the GE photographers with plenty of opportunities to capture him in a wide variety of settings and poses so as to give the magazines and Sunday supplements ever-fresh images of the engineer: Steinmetz bent over his work table, his laboratory apparatus, his makeshift desk in his canoe; Steinmetz meeting with Edison, Marconi, or Douglas Fairbanks; Steinmetz riding his bicycle, addressing a radio audience, or lying fully dressed on a rock in the middle of the Mohawk River. Playful photographs seem especially to have pleased the electrical engineer, who enjoyed playing in the darkroom and producing group photographs with multiple Steinmetzes.

Steinmetz is the subject of one of the short biographies that appear in John Dos Passos's novel *The 42nd Parallel,* the first in his *U.S.A.* trilogy. According to Dos Passos—a social historian and radical critic who was one of the major novelists of the "lost generation" that arose in the wake of World War I—"all his life Steinmetz was a piece of apparatus belonging to General Electric." GE publicists, said Dos Passos, "poured oily stories into the ears of the American public every Sunday," an apparent allusion to the weekly broadcast of Steinmetz's popular talks on the company's radio station.

Among the items issuing from GE's publicity department was a photograph of Steinmetz and Albert Einstein that was given wide distribution after the latter's 1921 Nobel Prize in physics. This famous photo appears to have prompted Dos Passos to write in his 1930 biographical sketch of Steinmetz that "all the reporters stood round while he and Einstein met face to face, but they couldn't catch what they said." According to John Jordan, writing on Steinmetz and the politics of efficiency in *Technology and Culture,* the journal of the Society for the History of Technology, the photo appeared in *Life* magazine as recently as 1965. I have seen it reproduced much more recently in articles that are often silent about Steinmetz's achievements and that do no more than touch on the scientist or the engineer, or on science or engineering.

Unlike many of the other photos of Steinmetz, which appear to

Einstein and Steinmetz

show him posed or caught in positions of mastery and authority
that deemphasized his congenital hunchback and stooped posture, the photo with Einstein seems to emphasize both. The father
of relativity stands tall, formal, and self-assured beside the short
and stooping, hand-in-pocket and smoking, rumpled and scowl-
ing figure who looks to be, ambivalently, both leaning toward the
famous scientist and pushing him back with an elbow. The incon-
gruity of the pair is accentuated by the juxtaposition of the hat-
ted Einstein's dark overcoat and the bareheaded Steinmetz's light
suit, as if the two were not even standing in the same weather.

It is ironic that, out of the myriad publicity photos of Steinmetz,
it is this one by which he is most frequently presented today. In
it he seems complimented only by his standing next to a Nobel
laureate, as if the engineer needed a scientist to lean upon. The
photograph's semiotics reinforce the stereotype of the engineer
vis-à-vis the scientist, of engineering versus science. In this picture the great engineer Steinmetz is reduced to a caricature of

himself; by implication, his achievement is diminished relative to that of the giant he stands beside.

The Einstein-Steinmetz photo is actually as false as are its implications for the relationship between science and engineering. The great scientist and the great engineer never did pose one-on-one, as they are presented in the photo. According to Jordan, who credits George Wise with pointing out to him the fabricated nature of the photograph, it was a creation of the overzealous publicity department at General Electric. Einstein and Steinmetz evidently were part of a group that attended a Radio Corporation of America–sponsored demonstration of transoceanic telegraph transmission at New Brunswick, New Jersey, in 1921, and a group photo was taken on that occasion. However, in that photo Einstein and Steinmetz were but two of about a score of people. Later, at GE, the group photo was altered to its bogus form and distributed after the announcement of Einstein's Nobel Prize.

Rather than by this doctored photo, which puts Steinmetz down in history as an uneasy inferior of Einstein, how much better it would be to remember him by his own multiple-exposure

Group photo from which Einstein and Steinmetz were extracted

photos that showed him to be a proud and playful genius in his own right. Among these latter are photos of lines of people interspersed among cloned Steinmetzes, boats with several bathing beauties each attended by a separate Steinmetz, and group portraits with many Steinmetzes but not a single Einstein. Whether he was given in the darkroom to dark thoughts about his single and childless life or whether they were just flights of fancy by which he relaxed, hundreds of such photos were among Steinmetz's belongings when he died.

In fact, Steinmetz's physical deformity and posture no more diminished the greatness of his engineering achievements than has Stephen Hawking's physical condition lessened his scientific achievements, or than Franklin D. Roosevelt's poliomyelitis downgraded his political achievements. Steinmetz seems to have come to grips with his misshapen body early in life, having determined not to marry and risk passing on to another generation the deformity he had inherited. A family photo of him upon arrival in New York shows the young man standing next to but not using a chair to disguise his condition; his 1909 portrait as president of the American Institute of Electrical Engineers demonstrates that he could appear quite dignified. In later GE publicity photos, the image of the famous Steinmetz is more often than not one of a lovable cigar-smoking eccentric leaning on a chair, stool, desk, or railing in such a way that the unnatural arch of his back is masked, akin to the way so many podiums hid the braced legs of Roosevelt.

Whether or not Steinmetz's physical deformity enhanced his love for humanity, as one writer has suggested, he did come to epitomize the engineer with a social conscience. Although Dos Passos wrote that "General Electric humored him [and] let him be a socialist and believe that human society could be improved the way you can improve a dynamo," there can be little doubt that Steinmetz's technical achievements were motivated by a sense of serving society.

Steinmetz definitely considered himself an engineer and not an inventor, given all the connotations of secrecy, entrepreneurship, and capitalistic competition associated with the latter. His contemporary Elbert Hubbard wrote: "Steinmetz resents being called an inventor. He says: 'I am only an engineer. My business is to construct engines that will transport an elemental form of energy into a million factories and homes, dividing this energy up into infinitesimal parts so it can be practically used to run sewing machines, to churn, to wash dishes and to do the dead lift and drudgery that otherwise would have to be done by human hands.'" As Steinmetz must surely have known, though, engineering is but institutionalized invention.

After a distinguished period of service on the Schenectady Board of Education and other involvement in local politics, in 1922 Steinmetz sought election to the post of New York State engineer and surveyor. His concerns for people extended to the environment, and he was troubled by the growing air and water pollution that accompanied the growth of cities and their power consumption. He proposed diverting the water that flowed over Niagara Falls (except on Sundays and holidays, in deference to tourists) and using its nine million horsepower to generate electricity, thereby saving fifty-four million tons of coal per year. His campaign was not successful, but when he died the next year, probably of heart failure hastened by the burden his deformity placed on his lungs, Steinmetz left a legacy of good works in both the technical and the social sphere. It is unfortunate that he is all too often remembered today only as the awkward figure beside Einstein. It is especially disturbing because this persistent image of an eclipse of the engineer by the scientist bears no relationship to actual events. The bogus photograph is but an artifact of imagemaking gone awry.

ALFRED NOBEL'S PRIZES

The Nobel prizes have their origins in a passage of fewer than three hundred words in the handwritten will that Alfred Bernhard Nobel executed by himself. Since it was prepared without the intervention of a lawyer, the will is remarkably free of legal jargon and easily read by a layperson. After specifying that a relatively modest proportion of his estate go to his heirs, the bachelor Nobel designated first that:

> The whole of my remaining realizable estate ... shall constitute a fund, the interest on which shall be annually distributed in the form of prizes to those who, during the preceding year, shall have conferred the greatest benefit on mankind.

It was thus the clear intention of Nobel that the prizes serve to give prompt recognition of achievement. The proximity of the award to the benefit to be recognized would seem alone to make it the kind characteristic of engineering or applied science rather than theoretical or basic science, and Nobel's principal intention appears to have been to reward technological products and processes at least as much as (if not to the actual exclusion of) abstract ideas or theories. Indeed, the common view of basic science, promulgated by basic scientists themselves, is that there is not necessarily any foreseeable or intended practical or material benefit to be derived from their research, although there may be some unintended practical consequences at some future time. Secondly in his will, Nobel spelled out that the prize money, which would come from interest on the fund comprising "safe securities" into which the realizable estate was to be invested, should be divided:

> ...into five equal parts, which shall be apportioned as follows: one part to the person who shall have made the most important discovery or invention within the field of physics; one part to the person who shall have made the most important chemical discovery or improvement; one part to the person who shall have made the most important discovery within the domain of physiology or medicine; one part to the person who shall have produced in the field of literature the most outstanding work of an idealistic tendency; and one part to the person who shall have done the most or the best work for fraternity between nations, for the abolition or reduction of standing armies and for the holding and promotion of peace congresses.

Nobel did not seem at all concerned with the professional credentials or status of those who would receive his prizes, or with professional fields as such. It was clearly the discovery or invention itself that was important to him, and whether it was the work of a professional or an amateur evidently did not matter. The terms "physics" and "chemistry" were apparently used in the will merely to distinguish areas of application rather than areas of research or professional affiliation.

Physics at the time of Nobel included the classical phenomena of mechanics, heat, light, electricity, magnetism, and the then new developments in X rays and other invisible rays. Chemistry, of course, dealt with phenomena that could not be explained by physical mechanisms alone. Even in the field of literature, Nobel appears to have specified his preference for work with a purpose beyond itself, what might be called "applied literature," such as the kind of storytelling that John Gardner called "moral fiction." Thus Nobel seems clearly not to have been interested in recognizing art for art's sake, and he can be assumed also not to have been interested in recognizing with a prize science for science's sake.

Thirdly, in his will Nobel specified what organizations or groups were to award the prizes:

> The prizes for physics and chemistry shall be awarded by the Swedish Academy of Sciences; that for physiological or medical works by the Caroline Institute in Stockholm; that for literature by the Academy in Stockholm; and that for champions of peace by a committee of five persons to be elected by the Norwegian Storting.

Although the cosmopolitan Nobel had lived as a citizen of the world, his family roots were in Sweden, which then included Norway. It will likely never be known whether Nobel gave any great thought to what organizations should award the prizes or whether he simply named, without consulting them, the organizations that he did for nostalgic reasons or because they seemed convenient and capable. However, it appears clear from the will that these organizations were charged only with awarding the prizes specified rather than with defining the prize categories. Unfortunately, since Nobel himself did not delimit his intended categories rigorously enough, the awarding institutions also became *de facto* the defining bodies for the prizes.

Finally, the will goes on to state that the prizes should be truly international, as was Nobel's practice of chemical engineering:

> It is my express wish that in awarding the prizes no considerations whatever shall be given to the nationality of the candidates, but that the most worthy shall receive the prize, whether he be a Scandinavian or not.

It is quite clear that Nobel's will in no way excluded engineering achievements from consideration for his prizes. Indeed, the will's first stipulation regarding the prize recipients was not that they be recognized or formal scientists, writers, or members of any profession, but rather that the products of their labor be worthy of recognition because they, "during the preceding year, shall have conferred the greatest benefit on mankind."

Almost seventy years before Nobel drafted his will, Thomas Tredgold proposed, as part of a request for a royal charter for the

Institution of Civil Engineers, a definition of engineering. Although couched in terms of "civil engineering," it applied then to all of engineering that was not directed to military ends. According to Tredgold's concept, it would seem that engineers are prime candidates for the prizes Nobel endowed:

> Civil engineering is the art of directing the great Sources of Power in Nature for the use and convenience of man; being that practical application of the most important principles of Natural Philosophy which has ... changed the aspect and state of affairs in the whole world.

The sources of power in nature were, of course, ultimately physical and chemical in the nineteenth century of Nobel, and the practical application of natural philosophy (i.e., science) is exactly what can lead to inventions and discoveries. Indeed, the language of Nobel's will looks almost to echo Tredgold's definition in that "benefit of mankind" could be equated with "convenience of man," and Nobel's explicit vision of the prizes as worldwide in scope is akin to Tredgold's implicit recognition that the ends of engineering also know no national boundaries.

Engineers are clearly engaged in making discoveries or inventions within the "fields" of physics and chemistry, and thus engineers should not at all be excluded *a priori* from Nobel's intentions. Furthermore, at the time of Nobel's will—in the closing years of the nineteenth century—the terms "science" and "scientific" were more commonly used to include engineering and technological pursuits than they are today. One need only read the "50, 100 and 150 Years Ago" page in the latest issue of *Scientific American* to see how totally integrated under the rubric of "science" were engineering and technology in days past. And a glance at an actual issue of *Scientific American* for 1895 makes the point even more emphatically.

In the year that Nobel wrote his will, *Scientific American* was identified on its cover as "a weekly journal of practical information, art, science, mechanics, chemistry, and manufactures." The cover illustration for a typical issue would show ship construction, a cable repair steamer, or some other new engineering achievement, and a typical sixteen-page tabloid-size issue would contain a full-page index of inventions for which letters patent were granted in the week of record. Many of the stories in *Scientific American* in 1895 were typically about what would be classified today as engineering or technology. In an issue dated within a week of Nobel's will, for example, there is a story about the effect of a Manhattan bank fire on structural steel, complete with an engineering drawing of beams and girders. Indeed, in the magazine's own contemporary words:

> Every number contains sixteen pages of useful information and a large number of original engravings of new inventions and discoveries, representing Engineering Works, Steam Machinery, New Inventions, Novelties in Mechanics, Manufactures, Chemistry, Electricity, Telegraphy, Photography, Architecture, Agriculture, Horticulture, Natural History, etc. Complete list of Patents each week.

What we would recognize more as "science" and research was by and large relegated to "a separate and distinct publication":

> THE SCIENTIFIC AMERICAN SUPPLEMENT ... presents the most recent papers by eminent writers in all the principal departments of Science and the Useful Arts, embracing Biology, Geology, Mineralogy, Natural History, Geography, Archaeology, Astronomy, Chemistry, Electricity, Light, Heat, Mechanical Engineering, Marine Engineering, Photography, Technology, Manufacturing Industries, Sanitary Engineering, Agriculture, Horticulture, Domestic Economy, Biography, Medicine, etc. ...
>
> *The most important Engineering Works,* Mechanisms, and Manufactures at home and abroad are illustrated and described in the SUPPLEMENT.

Thus it would seem that at the time that Nobel's will was drafted and translated into English, use of the terms "science" and "engineering" was at best convoluted. Two things can be stated incontrovertibly, however: (1) Things scientific definitely included the products of engineering; and (2) the public, as opposed to the institutional, status of engineering was at least as great as that of science. But when it came to awarding Nobel prizes, these conditions turned out not to work in favor of engineering.

The role of engineering was not always as it was at the end of the nineteenth century. Scientific societies were established centuries before Tredgold was called upon even to define engineering, and so science was firmly institutionalized as a distinct and well-defined field of endeavor long before Nobel's time. Engineering was certainly not even considered a legitimate field of study in universities like Oxford and Cambridge, or any universities for that matter, in the days of Newton, when the Royal Society was already flourishing. Specialized engineering as we know it today developed with the industrial revolution out of the arts and crafts of building and manufacturing steam engines, iron bridges, and other tangible products of technology. In fact, some of the first modern engineers did not *apply* science but rather *led* science. The science of thermodynamics may be viewed as an application of steam engines, and rational structural analysis as an application of bridge building. The view of scientific discovery as depending on the ingenious craftsmanship of instruments, and thus following technology, convincingly flies in the face of the conventional wisdom that technology is mere applied science.

Around the middle of the nineteenth century—when engineering had begun to form professional organizations such as the British Institution of Civil Engineers and the American Society of Civil Engineers, when the teaching of engineering was firmly established, albeit still primarily in technical schools as opposed to the long-established universities in which science was taught and practiced, and when engineering and technology were crucial in generating what has been described as the second industrial revolution with the growth of international trade and commerce of the kind in which Alfred Nobel himself was engaged—the rift between science and technology was growing.

By the turn of the century, when the first Nobel prizes were to be awarded, there existed a strong adversarial attitude on the part of science toward engineering, and science was unlikely to relinquish any advantage it might gain. The size, nature, and political if not professional advantages of the scientific, as opposed to the technical, community at the time, enabled it to act rather effectively and seize the opportunity presented by Nobel's will. In his acceptance speech, the first Nobel laureate in chemistry, J. H. van't Hoff, was to deplore the "drowning out of the gentle music of natural laws by the trumpet blasts of technical success" as he expressed clear approval of the awarding of prizes to discoveries of value in "pure" or theoretical science as well as in the "practical" or applied disciplines. Understanding how this state of affairs developed from Nobel's will, and how engineers have come to be in the predicament they find themselves in today, requires a look at how the expressed intentions of the will were translated into reality.

Alfred Bernhard Nobel was born in Stockholm on October 21, 1833. His father, Immanuel Nobel, was an engineer and inventor, and it was from him that Alfred was to learn engineering, as so many sons of his era did from their fathers. Also, like so many engineer-inventor-entrepreneurs of the time, the elder Nobel went bankrupt, prompting the family's move from Sweden to Russia in 1842. There success and prestige came to the elder Nobel, and Alfred was educated by private tutors, especially in chemistry and languages. There are discrepancies between official publications of the Nobel Foundation and standard reference works such as the *Encyclopaedia Britannica* about the details of Alfred Nobel's early career, but there is agreement on the fact that he began to work for his father's explosives concern as a mechanical and chemical engineer, and by 1867 he had received a patent for dynamite, which was to be largely responsible for his fortune.

Nobel reportedly acquired 355 patents during his life and apparently exploited them with vigor. He is said to have built some ninety factories and companies in twenty countries on five continents, and to have been a pioneer in the founding of multinational companies. While such activity intruded on his time for scientific research, which was the source of inventions for Nobel, it was necessary for him to realize rightly deserved financial rewards. Thus science and research for their own sake seem to have been far from an all-consuming love for Nobel.

Alfred Nobel dated his will November 27, 1895, about a year before he died. When it was opened, it surprised his relatives, who learned that they were left only about one of the 31 million Swedish krona (equivalent to about $4 million in late-twentieth-century dollars) that the estate was worth. Not only did there exist the threat that the family might contest the will, but there were at least three major problems faced by the executors: (1) jurisdiction over the will, since Nobel had left Sweden as a child and had never been a legal resident of any country; (2) the liquidation of property and the investment in "safe securities" as directed but unidentified in the will; and (3) the management of the fund and the development of rules for the distribution of the prizes.

The executors of the estate were also surprised by their role in having to deal with these problems. Apparently without their knowledge, Ragnar Sohlman, a chemical engineer like Nobel, and his assistant, and Rudolf Lilljeqvist, a mechanical engineer who had started the Electrochemical Company in 1895 for which Nobel had put up considerable capital, were named by Nobel as executors. The two engineers did not, however, play a significant role in deciding what the prizes themselves would become.

The executors had engaged Carl Lindhagen as attorney to the estate, and he took the lead in negotiating and drafting the procedures for the actual awarding of prizes. While Lindhagen, who was then beginning a political career, seems to have acted with the utmost integrity to see that the provisions of Nobel's will would be realized, his family connections and friendships made the Swedish scientific establishment especially accessible to him—and he to them. His uncle was an astronomer and was serving as secretary of the Royal Academy of Sciences. His father was one of the founders of the Stockholm Högskola, a private science-oriented institution, and Carl Lindhagen himself served as its secretary for ten years. Here he was closely associated with Professors Arrhenius and Pettersson, who would prove to be very influential charter members of the Nobel committees for physics and chemistry, respectively. Thus, from the beginning, representatives of the hard sciences and not engineering had the ear of the one individual who was in a privileged position to shape the institution of the Nobel prizes.

Lindhagen proceeded by drafting statutes to serve as the basis of negotiations between the executors and the institutions whose cooperation Nobel had more or less assumed. Lindhagen and Sohlman tried to clarify various aspects of selecting awardees, aspects that they anticipated might prove troublesome. In particular, definitions of "physics," "chemistry," and so forth were attempted, but in the final negotiations with the award-granting institutions these were by and large dropped. Spelling out more clearly what would constitute "importance" in the discoveries, inventions, and improvements that were to be the subjects of the prizes also proved to be difficult for the scientist-negotiators of the institutions to pin down. They were accustomed to scientific societies conferring honors based principally on past—sometimes long past—performance, especially that reported in published works. In the final statutes only published works were eligible for a Nobel prize, and such a provision was certainly not in the best interests of engineers, who tend to realize their ideas in machines and structures rather than in words on paper.

In the end, the statutes that were agreed upon by all parties set up an organizational machinery but not rules concerning the choice of works for which or for the delineation of fields in which the prizes would ultimately be awarded. The Code of Statutes of the Nobel Foundation, finally promulgated by King Oscar II of Sweden and Norway on June 29, 1900, began: "The Nobel Foundation is based upon the last Will and Testament of Dr. Alfred Bernhard Nobel, Engineer."

Although the statutes go on to quote the relevant paragraph of the will and to refer repeatedly to it, there is much more attention paid to the mechanics of administering the prizes and the composition, rights, and rewards of members of the Nobel Foundation than there is to an explication of the will proper. And once established, the Nobel Foundation became and has remained essentially self-perpetuating and thus difficult to change in composition or in mind. Furthermore, according to Section 10 of the Statutes: "Against the decision of the adjudicators in making their award no protest can be lodged. If differences of opinion have occurred they shall not appear in the minutes of the proceedings, nor be in any other way made public."

Thus no degree of public accountability of the adjudicators, who were empowered to establish Nobel institutes, was encouraged. As long as the institutes maintained a semblance of order and rationality among their narrowly defined and self-defined peer groups, especially in the fields of physics, chemistry, physiology, and medicine, there could be little realistic hope of outsiders to those fields becoming insiders on the Nobel committees. Of the members of the physics and chemistry committees for the first fifteen years of the prizes, only one member, A. G. Ekstrand, who served from 1913 to 1924, is listed as an "engineer," and he does not appear to have played any major role in formative policy-making. To this day, according to the Nobel Statutes, the right to submit proposals—that is, nominations—for the award of science prizes is enjoyed only by a select group that is unlikely to include many engineers.

Clearly the politics of the Nobel prizes and how they were created in practice from the rather sketchy will of Alfred Nobel work against the interests of engineering. Had Nobel created a more explicit and rigid document, he might have made clear and unambiguous what he meant by "invention," "discovery," and other terms in his will, and he might have spelled out clearly whether engineers were eligible for his prizes or to sit on selection committees. As it turns out, the very imprecision of his will enabled those who were to gain control of the nomination process to shape it to their own ends. One publication of the Nobel Foundation even admits that "had the will been written in strict legal style it would probably have been impossible to adapt it to contemporary times." How it is adapted of course depends upon who is in control of the adaptation.

Technology is included among research specialties that comprise areas for prospective prizes, but Nobel prizes in this category have been few and far between. The issue of awarding prizes to engineers for contributions to engineering, as opposed to science, was a matter of informed public comment in the early years of the prizes, and in 1908 an editorial titled "The Purpose of the Nobel Prizes" appeared in the *New York Tribune*. It read, in part:

> One of the most remarkable features of the [Nobel Foundation] is that it ignores the profession in which Nobel was himself trained.... Nobel was a mechanical engineer; at least early in his life. It might have been supposed that he would have had a lively appreciation of the service rendered to mankind by devotees of that branch of science in the past; for instance, in developing improved means of transportation and appliances for the operations of manufacture. He could not have imagined that that chapter of history had closed. For a man with his particular experience to show a preference for other departments of study must be regarded as more or less strange.

Apparently there was some pressure on the foundation from engineers in the early years, and in 1912 the recommendation of the physics committee was rejected in favor of the Swedish inventor Nils Gustaf Dalén "for his invention of automatic regulators for use in conjunction with gas accumulators for illuminating lighthouses and buoys." But this, along with the prize in 1909 to Guglielmo Marconi and Ferdinand Braun "in recognition of their contributions to the development of wireless telegraphy," was an anomaly, and such entries are rare on the lists of Nobel prizes. The invention of the airplane, for example, went unrecognized by the Nobel Committee, apparently because the loss of life it could potentially cause was considered to outweigh its possible benefit to mankind. Such a concern is ironic considering the destructive power of dynamite, which endowed the prizes, and the consequences of discoveries in particle and atomic physics, which have captured so many Nobel prizes.

So it is understandable why in the 1980s engineering groups were advocating a separate Nobel prize. But appeals for a new prize category were almost certainly doomed to failure, for every special-interest group would have liked to have its own separate category, and the Nobel Foundation was not interested in diluting the impact of the rare prizes. In 1969 a single award category was added to those in Nobel's will, but it is technically not a Nobel prize but rather "The Bank of Sweden Prize in Economic Sciences in Memory of Alfred Nobel." However, controversy seems to be generated by any prizes associated with the Nobel name, and some economists have actually called for the abolition of the economics prize because of their perception that there is a paucity of outstanding candidates and that evaluation of excellence in the field is so charged with political and social values that objectivity is elusive.

The addition of the economics prize was unique, and Stig Ramel, speaking as executive director of the Nobel Foundation in 1983, stated that the foundation had decided a few years earlier that no further prizes would be established. Thus the appeal from engineers was doomed from the start, and their bid for a separate engineering prize was rejected formally in 1986.

Two years later, the National Academy of Engineering established the Charles Stark Draper Prize, which has since been described as the "Engineering Nobel Prize." Backed by an endowment from the Charles Stark Draper Laboratories of Cambridge, Massachusetts, as a memorial to their namesake, the Draper Prize is awarded biennially "to recognize individuals whose outstanding engineering achievements have contributed to the wellbeing of humanity." The first Draper prizes were awarded in 1989 to the two engineers, Jack S. Kilby and Robert N. Noyce, who independently invented and developed the integrated circuit. Subsequent prizes have recognized engineers responsible for the jet engine, the computer language FORTRAN, and communications satellite technology. But there has been relatively little public recognition of these awards compared to the annual media frenzy associated with the announcements of the Nobel prizes. The apparent desire of Alfred Nobel to recognize engineering accomplishments has turned out ironically to have had the opposite effect.

HENRY MARTYN ROBERT

Few people who attend meetings, whether of social organizations, citizen groups, or university faculty, have failed to hear Robert's Rules of Order invoked at one time or another. More often than not, however, the rules are more easily appealed to than remembered, and for all the good intentions of those assembled, few meetings proceed in strict adherence to Robert's Rules. In fact, perhaps only the engineer who laid down the rules in the first place is less known than some of the finer points of parliamentary procedure now associated with his name.

Henry Martyn Robert was born in 1837 on the family plantation near the small town of Robertville, South Carolina, about forty miles up the Savannah River from Savannah, Georgia, near the state border. He was the middle of seven children born to the Reverend Joseph Thomas Robert, a Baptist clergyman, and Adeline Elizabeth Robert, whose father was a military man and whose brother was to become Confederate Army general Alexander R. Lawton. Among Henry Robert's paternal ancestors was Pierre Robert, one of South Carolina's first settlers in the seventeenth century and its first Huguenot pastor.

In 1853, before the Union was split apart, Henry Robert was appointed to the U.S. Military Academy, from which he graduated in 1857. He excelled in mathematics at West Point, and while still a cadet acted on one occasion as assistant to the professor of mathematics. For a year after graduation he served as an assistant professor of practical military engineering, and concurrently taught natural philosophy and astronomy at the academy.

In 1858, under assignment to the Corps of Engineers, he was ordered to duty in Washington Territory, in the northwestern part of the young country, to take command of engineering operations in a campaign against the Indians. Since transcontinental rail routes were not yet developed, Robert and his men took the then familiar sea-land-sea route from the East Coast to the West Coast, which meant crossing the Isthmus of Panama, where Robert contracted "Panama fever" (the residual effects of which would keep him from seeing battle).

Out West, Robert began exploring wagon and military routes. When a dispute arose over the U.S.–Canadian border, however, he and other troops in the area found themselves confined to San Juan de Fuca Island, on the contested border off the southeastern tip of Vancouver Island, and here they constructed defenses against the British in anticipation of a battle. Robert did not engage in any hostilities in the Northwest, however, for with the Civil War looming he was ordered back East to take charge of the defenses of the city of Washington. The lingering effects of his tropical disease led to his reassignment to work on defenses in more northern areas, first to Philadelphia and then to Bedford, Massachusetts. By the end of the war Robert had reached the rank of captain and was again assigned to West Point, where he rejoined the Department of Practical Military Engineering. After a term as treasurer of the academy, he was promoted to major and assigned to the West Coast once again, this time to serve in San Francisco as chief engineer of the military division of the Pacific.

Robert's career as a military engineer subsequently involved a considerable amount of moving about in the developing United States, and he served as superintending engineer of river and harbor improvements, as well as of military defenses, in such areas as Oregon and Washington; on Lakes Superior, Michigan, Erie, Ontario, and Champlain; on the St. Lawrence and Tennessee rivers; and in Delaware Bay, Long Island Sound, and New York Harbor. Such a peripatetic career was the norm for engineers of the day, both military and civilian. While the military engineers were maintaining navigation channels and harbors, the civilian engineers were more often than not laying out railroads and building bridges across many of the same waterways that the military wished to keep clear. This led to not a few disputes over the rights of water navigation versus land travel between military and civilian engineers, which incidentally pushed bridge engineering to develop longer and higher spans to give ample navigation clearance to steamboats and sailing ships.

By 1895 Robert had been promoted to colonel and was placed in charge of the Southwest Division, which included eleven engineering districts ranging from Pittsburgh to Galveston, Texas, where he did his most significant engineering work. About five years earlier, as a member of a board of engineers appointed to recommend where on the western coast of the Gulf of Mexico a port should be developed to handle growing volumes of cotton and other shipping, Robert had advocated the construction of jetties defining a wide channel that would maintain itself without dredging. The leadership of the Corps of Engineers had included many bitter opponents to such a scheme when it was proposed by James Buchanan Eads in the 1860s in connection with maintaining a channel at the mouth of the Mississippi, but by the time Robert was arguing for a similar treatment at the port of Galveston, Eads's work in the New Orleans area had come to be recognized for the success that it was. In time, after the installation of the jetties in Galveston, the sandbar that had blocked the entrance to the bay was moved steadily into deeper water by the increased water velocity between the jetties and was diminished in size, just as Robert had predicted it would be.

In 1901, as a reward for his service, Robert was promoted to the rank of brigadier general and was made chief of engineers, but mandatory retirement laws allowed him to hold that position for only three days. Like many a retired military engineer, Robert continued to work as a consulting engineer. His prior work in establishing the Port of Galveston led naturally to his being kept on retainer by that city. After a tidal wave struck the city and drowned six thousand people during a hurricane in 1900, Robert chaired a board of engineers that was asked to come up with measures to prevent future such disasters. A seawall was recommended, and one was built that effectively protected the city during subsequent storms in 1909 and 1915, after each of which Robert was called upon to serve Galveston further. A granite monument erected atop the seawall bore the names of the Board of Engineers responsible for it: General Robert; Alfred Noble, then president of the American Society of Civil Engineers; and H. C. Ripley, former city engineer of Galveston.

*Brigadier-General Henry
Martyn Robert*

Like many engineers, especially those in charge of significant projects, Robert was expected to write reports and monographs. In 1881 he published *The Water-Jet as an Aid to Engineering Construction,* and he also compiled indexes to engineering reports of the Corps of Engineers on river and harbor improvements during the important period of development following the Civil War. But while pursuing a full and distinguished career as an engineer,

Robert pursued with equal vigor an avocation that led to his name becoming inextricably associated with parliamentary procedure.

Robert's introduction to the problem of order is said to have come when, as a young engineer assigned to duty in Massachusetts, he was asked to preside at a small but unruly church meeting. At the time he knew nothing of parliamentary procedure, and he later wrote of the experience, "My embarrassment was supreme." He continued, "I plunged in, trusting to Providence that the assembly would behave itself. But with the plunge went the determination that I would never attend another meeting until I knew something of... parliamentary law." Robert believed that knowledge of that law was "a part of the necessary equipment of every educated man," and he was determined to learn it.

He found little written on the subject, however, and what he later discovered showed little consistency or conformity with practice. Luther S. Cushing, a noted jurist, had published in 1845 a *Manual of Parliamentary Procedure,* and it was then to popular assemblies what Thomas Jefferson's *Digest of the Rules of Congress* was to legislative bodies. Apparently neither of these works was known, readily available, or thought highly of by Robert when he began his own rules. Later, upon being reassigned to San Francisco, where he encountered people from all parts of the country, Robert found further that there was a great difference of opinion as to what constituted the proper way to conduct a meeting. Robert first looked to Cushing's *Manual* as an arbiter, but he found it inadequate and typographically unsuited for use as a ready reference.

As he would later outline in the introduction to his own rules of order, parliamentary law "refers originally to the customs and rules for conducting business in the English Parliament." These unwritten practices were modified for use by the U.S. Congress, but there were important differences between practice in the House of Representatives and in the Senate. Robert noted, for example, that with regard to the order of precedence of motions, the previous question was admitted in the House but not in the Senate. Also, with the growing agenda of a growing bipartisan Congress, parliamentary procedure allowed the majority to limit and even suppress debate in Congress, something he felt should not be permitted in a deliberative body. *Rules of Order* was thus written to adapt parliamentary law "in its details to the use of ordinary societies." Furthermore, according to Robert,

> The object of Rules of Order is to assist an assembly to accomplish in the best possible manner the work for which it was designed. To do this it is necessary to restrain the individual somewhat, as the right of an individual, in any community, to do what he pleases, is incompatible with the interests of the whole. Where there is no law, but every man does what is right in his own eyes, there is the least of real liberty.

Robert's interest in the democratic process is also demonstrated in another widely quoted statement of his, in which may be read a hint of the great intersectional struggle of his lifetime, the Civil War:

> The great lesson for democracies to learn is for the majority to give to the minority a full, free opportunity to present their side of the case, and then for the minority, having failed to win a majority to their views, gracefully to submit and to recognize the action as that of the entire organization, and cheerfully to assist in carrying it out, until they can secure its repeal.

Robert's first attempt to set down more democratic rules of order resulted in "a fifteen-page manual for the guidance of himself and his friends in conducting the deliberative work of charitable, social and civic organizations in which they might be interested," according to E. J. Mehren, editor of *Engineering News-Record*, who in 1920 published in that journal the appreciation of Henry M. Robert that now forms a principal source of information on the engineer and his pursuits. Mehren recalled first meeting Robert while sitting on the piazza of the Hotel Galvez in

Galveston. The general was introduced as the engineer of the sea-wall, the story of which he proceeded to recount in great detail for the engineering journalist. Only after an hour of harbor history did the conversation turn to publishing, at which time Robert revealed that he had written a "little book" whose story was equally fascinating.

After Robert had been transferred to Milwaukee, he determined to use the leisure provided by the severe winter of 1874, which tied up army activities on Lake Michigan, to work on an authoritative manual of order. Taking the problem to be one whose solution could be approached by systematic engineering analysis, Robert "analyzed each motion and each step of procedure carefully" and tested his approach to them against principles that would "guarantee the deliberative character of the assembly and safeguard the rights of individual members." When later that year Robert completed the manuscript for a book on parliamentary law, "one of the leading Eastern publishing houses," D. Appleton and Company in New York, rejected the opportunity to publish it because there was believed to be only a "small demand for such a book." The determined Robert decided to publish the book himself, buying the paper to print it on and selecting the type "in order that the various sections and subdivisions might be thrown in such relief as to appeal to the reader." The production process was an arduous one, since Robert's duties with the Corps of Engineers allowed him to proofread text only on occasion. Because of this, just sixteen pages were printed at a time, so as not to tie up too much of the printer's type. But Robert was determined to attend carefully to every detail, and he even selected the familiar pocket size of the volume so as to make it convenient for carrying to meetings.

In its original form, *Robert's Rules of Order* was just that—rules, as is the first part of the work today. When his wife, Helen Marie Robert, pointed out that this bare declaration of rules would be "quite baffling to one who had never had parliamentary experience," Robert added a second part to his manual. This part of the book instructs the reader on what must be done before calling "a meeting that is not one of an organized society." It also deals with such practical matters as how long to wait after the scheduled time before starting a meeting and how to select a chairman and secretary before the body has elected either. Robert also added, after a member of a religious society contested his ejection from a meeting, a third part, describing the legal rights of assemblies and procedures for regulating the behavior of their members. This now is incorporated into the second part of the rules.

Even after Robert had printed and bound his book, whose full original title was *Pocket Manual of Rules of Order for Deliberative Assemblies,* he could interest a Chicago publisher, S. C. Griggs Company, to distribute it only after Robert offered to give away a thousand free review and promotional copies to legislators, professors, legal authorities, and the like. This left three thousand copies from the first printing, which Robert estimated would last for two years, thus giving him ample time to collect feedback and prepare a revised edition. Ever the engineer, Robert knew that a book, like a harbor work, would have certain weaknesses in its design revealed only upon being tested in use, and so he anticipated "criticisms which would be useful in revising the book." Thus "I had not had electrotype plates made because I wished to revise the book before the type were set for the plates."

In fact, the initial reception of the book was so positive that it was sold out within four months of its publication in February 1876. The printing plates that were eventually made wore out in 1893, at which time further revisions were incorporated before new plates were made. Organizations by this time had begun to make explicit reference in their bylaws to *Robert's Rules of Order* as the final authority on any matter not explicitly covered by the organization's own rules. By 1914 more than five hundred thousand copies of *Robert's Rules of Order,* in various printings, had been sold. By 1915 "the constant inquiries from all sections of the country for information" not contained in the book had prompted the preparation of a new and enlarged edition said to comprise more than 75 percent new material, thus warranting a new copyright. To avoid confusion with the old edition, the new one, published by Scott, Foresman, and Company, appeared under the title *Robert's Rules of Order, Revised.*

Henry Robert spent the last years of his life in Hornell, New York, near the home of his second wife, Isabel Livingston, where at the age of eighty-three he oversaw the remodeling of an old house in which he continued to write. He died in Hornell in 1923 and was buried in Arlington National Cemetery, but his work has been a unique legacy for Robert's descendants. By incorporating changes left by the author in his own marked copy of *Rules of Order,* and by updating such details as references to Congress as it has changed its own rules, Robert's heirs have continued to revise the work to keep it current and useful; their names have been added to the title page as editors and assistants, as well as continuing to appear on the copyright page as beneficiaries of Robert's resolve and estate. *Robert's Rules of Order, Newly Revised* appeared in 1970 as the seventh edition of the manual, and in 1990 the ninth edition was published. Like the foundation of a well-conceived jetty or seawall, however, the defining concept of the engineer's original edition of *Robert's Rules* continues to give the work as evolved its lasting character and value. And the name of the generally forgotten engineer has become one with his work.

JAMES NASMYTH

Few engineers have been more explicit about the importance of drawing to their professional practice than was James Nasmyth, who was born in Edinburgh in 1808. He was the son of the painter Alexander Nasmyth, who has been called the "father of Scottish landscape art" and who was an excellent mechanic in his own right. Besides a painting room, the older Nasmyth also had a well-equipped workroom, in which he relaxed away from his easel among shelves of "artistic and ingenious mechanical objects, nearly all of which were the production of his own hands." To this room were admitted only his most intimate friends and those acquaintances who could appreciate the mechanisms arranged as neatly as the illustrations in a catalog of designs.

Alexander Nasmyth the mechanic is credited by his son with inventing in 1794 a "bow-and-string bridge," what we today would call a tied arch, the first one of which was erected over a deep ravine on the island of St. Helena in the South Atlantic. Although the strength of the lightweight iron bridge was questioned at first, it was used with the utmost confidence after it did not suffer the least bit of damage under the herd of wild oxen that rushed across it in a proof test. The older Nasmyth is also credited with inventing a method of riveting without loud hammer blows. He is reported to have used the jaws of his bench vise to squeeze the heads on hot rivets, thus repairing an iron stove "in the most perfect silence" on a Sunday morning. This "Sunday Rivet" method produced a tight and reliable joint, and thus it came to be widely used in making iron boilers and building ships. The technique was employed in riveting the wrought-iron plates of the famous Britannia Bridge, a tubular structure that spanned the Menai Strait in northwestern Wales.

Sunday engineering by a nineteenth-century artist might at first thought seem to be as unusual as the Sunday painting of a late-twentieth-century engineer. But that is merely the conventional wisdom. In fact, the kind of engineering done by Alexander Nasmyth could be and was done by many who had the time, facilities, and inclination to do so, and many of those who did so spent more than just their Sundays at it. An artist or other imaginative and creative individual could do engineering without formal training, just as an engineer can paint without art lessons. And some were better at engineering than others, just as some paint better than others. What appeared to distinguish the accomplished from the not-so-accomplished was an appreciation of what had been and was to be achieved, whether they were ingenious mechanical devices or well-executed paintings.

It was in his father's workroom that James Nasmyth first began to handle mechanical tools, an experience he called "my primary technical school—the very foreground of my life." Nasmyth did not much enjoy high school, especially disliking Latin. After high school he continued his studies in private classes, and Euclid stimulated him intellectually. But what really captured Nasmyth's imagination was being able to take part in the activity at a local iron foundry:

> Nothing could be more agreeable to my tastes, for there I saw how iron castings were made. Mill-work and steam-engines were required there, and I could see the way in which power was produced and communicated. To me it was a most instructive school of practical mechanics. Although I was only thirteen at the time I used to lend a hand, in which hearty zeal made up for want of strength. I look back to these

days, especially to the Saturday afternoons spent in the workshops of this admirably conducted iron-foundry, as a most important part of my education as a mechanical engineer. I did not *read* about such things, for words were of little use....

Young Nasmyth also experimented with chemicals. He and a friend made as many as possible of their own chemicals, and Nasmyth considered this experience indispensable in teaching him about materials. And in his autobiography, edited (if not ghostwritten) by Samuel Smiles and published posthumously in 1883, Nasmyth would lament the degree to which technical education toward the end of the nineteenth century had become removed from the workshop experience of his youth. He deplored the practice of parents buying their young sons ready-made models of ships and steam engines and then "after paying vast sums" to have the "young imposters" apprenticed to engineering firms, finding them learning only "glove-wearing and cigar-smoking!" Nasmyth then preached:

The truth is, that the eyes and the fingers—*the bare fingers*—are the two principal inlets to sound practical instruction. They are the chief sources of trustworthy knowledge in all the materials and operations which the engineer has to deal with. No *book* knowledge can avail for that purpose. The nature and properties of the materials must come in through the finger-ends; hence I have no faith in young engineers who are addicted to wearing gloves.

Thus Nasmyth, who was an engineer of the transition from the learning-by-doing school of the first part of the nineteenth century to the learning-by-learning school of the latter part, was saddened by the passing of engineering, which he defined as nothing but the application of common sense to materials, from an experience of the hands to one of the head. But we are, as he was, getting ahead of his life's story.

Among the advantages young James Nasmyth had was the opportunity to practice the art of drawing as taught by his father:

James Nasmyth

He taught me to sketch with exactness every object, whether natural or artificial, so as to enable the hand to accurately reproduce what the eye had seen.... He would throw down at random a number of bricks, or pieces of wood representing them, and set me to copy their forms, their proportions, their lights and shadows....

My father was an enthusiast in praise of this *graphic language*, and I have followed his example; in fact, it formed a principal part of my own education. It gave me the power of recording observations with a few graphic strokes of the pencil, and far surpassed in expression any number of mere words. This graphic eloquence is one of the highest gifts in conveying clear and correct ideas as to the form of objects, whether they be those of a simple and familiar kind, or of some form of mechanical construction, or of the details of a fine building, or the characteristic features of a wide-stretching landscape. This accomplishment of accurate drawing...served me many a good turn in future years with reference to the engineering work which became the business of my life.

Before he did get on with the business of his life, however, Nasmyth attended the first classes of the Edinburgh School of Arts, founded in 1821 as a technical college after the model of mechanics institutes that had begun in Glasgow. There Professor John Anderson, who held the chair of natural philosophy at Glasgow University, had lectured in the evenings to classes that included townspeople "of almost every rank, age and employment," and promoted ideas leading to giving instruction in the principles underlying the activities of workingmen and mechanics. Nasmyth attended the school until 1826, taking courses in chemistry, mechanical philosophy, geometry, and mathematics. He made sectional and working models of steam engines, not only for lecture demonstrations at the school but also for lectures on natural philosophy at the university.

In the course of his study of engines made by various firms, Nasmyth learned that the London machine shop of Henry Maudsley was "the very centre and climax of all that was excellent in mechanical workmanship." Nasmyth desired to be taken on there as an apprentice, but he also wondered if it was possible. He wrote, "I was aware that my father had not the means of paying the large premium required for placing me as an apprentice at Maudsley's firm. I was also informed that Maudsley had ceased to take pupils."

Apparently no more pupils were being taken in the machine shop because apprentices had been "coming in gloves," keeping irregular hours, and generally disrupting the business of the place. Still, Nasmyth hoped that by showing Maudsley examples of drawings and models, an exception might be made, and so the young hopeful proceeded to make a "most complete working model of a high pressure engine." Also,

In a like manner I executed several specimens of my ability as a mechanical draughtsman, for I knew that Maudsley would thoroughly understand my ability to work after a plan. Mechanical drawing is the alphabet of the engineer. Without this the workman is merely "a hand." With it he indicates the possession of "a head."

Nasmyth's enthusiasm appealed to Maudsley, and when he saw the quality of the young man's work he was taken on not as an apprentice but as an assistant in the master's private workshop. They worked together until Maudsley died in 1831, when Nasmyth struck out on his own, and not in a small way, for he began making such things as machine tools and locomotives.

In 1839 Nasmyth received a letter that led to his master work. He knew that the *Great Britain* was being built in the Bristol shipyards, and, it being the largest ship made at the time, that a very large iron shaft was required, but Nasmyth had not realized until the letter informed him so, that there was no forge hammer in England or Scotland powerful enough to do the job. So he thought about what it was about existing forge hammers that limited their capacity—what it was about them that caused them to fail to be more powerful—and he understood that the design problem was to contrive some method of repeatedly raising and lowering a heavy iron block with sufficient accuracy to accomplish the job. Then, according to Nasmyth,

> Following out this idea, I got out my "scheme book," on the pages of which I generally *thought out*, with the aid of pen and pencil, such mechanical adaptations as I had conceived in my mind, and was thereby enabled to render them visible. I then rapidly sketched out my steam-hammer, having it all clearly before me in my mind's eye.

Thus within a half hour of being told of the problem of forging a very large shaft, Nasmyth had sketched out his solution. The idea of the steam hammer was communicated to, among others, Isambard Kingdom Brunel, the engineer-in-chief of the great steamship, and the invention was heartily approved. Nasmyth then gave the shipbuilders permission to divulge his concept to any iron forge that might wish to erect such a steam hammer, under the condition that his own firm would be allowed to supply it.

In the meantime, however, a 237-ton steamship driven not by paddle wheels but by a screw propeller entered Bristol Harbor, where the *Great Britain* was under construction. The *Archimedes* was the first successful screw steamer in the world, and when Brunel saw it he immediately suspended work on his own ship's paddle-wheel propulsion system and contracted to have use of the *Archimedes* for six months of sea trials. The new means of driving a ship proved so superior that Brunel determined to re-design the *Great Britain* to be a screw-driven ship. When the Admiralty remained skeptical about Brunel's preference, he arranged for a tug-of-war between two sloops, the *Alecto* and the *Rattler,* identical but for their means of propulsion, the former being driven by then-conventional paddle wheels and the latter with the new system. When both vessels went full steam ahead, the *Rattler* towed her opponent away at a speed approaching three knots.

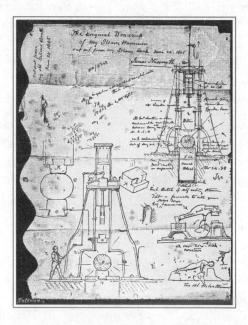

Page from Nasmyth's "scheme book" showing
his concept of the steam hammer

The *Great Britain*'s paddle-wheel concept, to which a large iron forging was central, was abandoned and replaced by a screw propulsion system, so there was no longer an urgent need in Britain for Nasmyth's steam hammer. Furthermore, the year was 1840 and there was a widespread depression and the iron trade had fallen on bad times. Appealing as the steam hammer was in principle, there were not sufficient orders for any ironworks to invest in such a major piece of new equipment, and it existed as a sketch only.

However, Nasmyth continued to do a good business selling machine tools to foreign businesses, and he freely shared his new ideas with visitors from abroad, opening his scheme book for their inspection. One visitor, a proprietor of a French iron works, upon being shown Nasmyth's scheme book in his absence, made a sketch of the idea and thus took a copy of the steam hammer concept back to France. It was only by chance, on a visit abroad two years hence, that Nasmyth saw his steam hammer sketch realized. It had been built and was being used in an ironworks in Creuzot, France. The inventor had mixed feelings of pleasure and concern when he was shown by M. Bourdon, the forge's mechanical manager, the great steam hammer:

> [T]here it was in truth—a thumping child of my brain! Until then it had only existed in my scheme book; and yet it had often and often been before my mind's eye in full action. On inspecting the steam-hammer I found that Bourdon had omitted some important details, which had led to a few mishaps, especially with respect to the frequent breaking of the piston-rod at its junction with the hammer-block.... I sketched for him, then and there, in full size, on a board, the elastic packing under the end of the piston-rod, which acted, as I told him, like the cartilage between the bones of the vertebrae, preventing the destructive effects of violent jars. I also communicated to him a few other important details, which he had missed in his hasty inspection of my design. Indeed, I felt great pleasure in doing so....

I had not yet taken out a patent for the steam-hammer. The reason was this: the cost of a patent, at the time I invented it, was little short of 500 pounds sterling, all expenses included. My partner was unwilling to lay out so large a sum upon an invention for which there seemed to be so little demand at that time, and I myself had the whole of my capital embarked in the concern. . . .

I now became alarmed, and feared lest I should lose the benefits of my invention. . . . [M]y brother-in-law . . . had . . . many times cautioned me against "publishing" [the steam-hammer's] advantages so widely, without having first protected it by a patent. He was, therefore, quite ready to come to my assistance. He helped me with the necessary money [and] the patent was secured in June, 1842, or less than two months after my return from France.

Soon the mania for expanding the railroads had struck Britain, and with it and other developments the iron trade recovered from its depression. Nasmyth's steam hammer, which also became indispensable for forging sound anchors for the large steamships of the time, was widely adopted, not only at home but also abroad, including in America. The mounting orders brought substantial financial rewards to Nasmyth and his brother-in-law.

An example of his steam hammer was on display at the Great Exhibition of 1851 in London, and in his autobiography Nasmyth described the machine with all the warmth that others had reserved for the Crystal Palace itself:

The machine combined great power with gentleness. The hammer could be made to give so gentle a blow as to crack the end of an egg placed in a wine-glass on the anvil, while the next blow would shake the parish, or be instantly arrested in its descent midway.

His invention so pleased James Nasmyth that the steam hammer in use in his Manchester foundry became the subject of a striking oil painting that he executed himself. Indeed, according to one contemporary viewer of his paintings, Nasmyth "would have won fame for himself as an artist . . . had not science and mechanical invention claimed him for their own." Thus it should have come as no surprise that it was not only for his steam hammer that Nasmyth was awarded a medal at the Great Exhibition. One prize that especially pleased him was for his drawings of the moon's surface. Astronomy was one of Nasmyth's many, many hobbies, and his moon map was so admired by the prince consort when he saw it at the Crystal Palace that he asked Nasmyth to show Queen Victoria his astronomical drawings when she visited Manchester after the close of the Great Exhibition.

As the steam hammer and his other interests brought visitors to Nasmyth, so his position and natural curiosity also took him abroad on many occasions. While the Great Exhibition brought the works of industry of all nations to London for all to see, up until that time one generally had to travel to various nations and workshops separately to observe with one's own eyes what ingenious inventions were new and different. And the practice did not begin in the nineteenth century.

According to Eugene Ferguson, "intellectual currents of the sixteenth and seventeenth centuries encouraged the visual, nonverbal study of machines. A few humanists foresaw an end to learning, as disputations on ancient texts loaded the mind with wordy formulas that had little substance." In 1580 a noted French potter, Bernard Palissy, encouraged scholars to come to his workshop, where they might see for themselves that "the theories of many philosophers, even the most ancient and famous ones, are erroneous in many points." In his *Dialogues* of 1638 Galileo encouraged visiting the shops of Renaissance mechanics to observe the work of "first rank men" and learn firsthand of failures of designs that might initially have "seemed to all an excellent idea." By 1669 Isaac Newton was advising a young friend about what it might be to his best advantage to observe in his upcoming travels on the Continent: "Trades & Arts," "Mechanisme & manner of guiding ships," "circumstances of mining & of extracting metals or mineralls out of their oare and refining them."

In the 1650s the English diarist John Evelyn, the political economist William Petty, and others began compiling "A History of
Arts Illiberal and Mechanical," and they soon recognized the need
for illustrations, because, according to Petty, "bare words being
not sufficient, all instruments and tools must be pictured, and
colours added, when the descriptions cannot be made intelligible
without them." One fellow of the Royal Society of London even
thought that a history of the trades would encourage greater inventions than the craftsmen were capable of, and in a statement
anticipating the rift that would eventually grow between modern
scientists and engineers, he claimed that "a large and unbounded
Mind is likely to be the Author of greater Productions, than the
calm, obscure, and fetter'd Endeavours of the Mechanics themselves."

In France a similar effort was being promoted by 1675, but
progress came slowly, and writing did not begin until 1704. When
Denis Diderot began planning *L'Encyclopédie,* he drew on the earlier English effort, including some 150 drawings and engravings
that had been prepared for it. The published *Encyclopédie* contained nearly three thousand full-page plates, and it has been estimated that about half of the illustrations pertain directly to and
clarify the written descriptions of the crafts and trades. Diderot's
ultimate objectives were to produce an illustrated series on arts
and crafts "that would induce Artists to read, Philosophers to
think usefully, and men of power to make effective use of their
authority."

It is out of such a growing realization of the value of the arts
and crafts to civilization and culture that modern engineering
started to rise with determination in the eighteenth century. And
as the industrial revolution gained momentum and technology
began to change at a much faster pace than any encyclopedia,
whether illustrated or not, could keep up with, there was still no
substitute for travel. And it was not only artists who came back
from their travels with sketchbooks full of memories and images;
the emerging civil and mechanical engineer epitomized by James
Nasmyth came back with drawings for new machines and
structures.

Nasmyth traveled widely, and always with his sketchbook, even after the introduction of photography threatened the practice of sketching what one had seen with one's own eyes. On one occasion in Scandinavia, he even drew in his sketchbook the food and drink he wanted in "the universal language of the pencil" when he could not speak the local language. And sometimes travel would be more for commercial than for research purposes. In 1842, for example, he traveled to Nuremberg to confer with the directors of the Nuremberg & Munich Railroad, who needed locomotives for their line. He and his partner had gone in person, since the transaction was an "important and extensive" one, according to Nasmyth, one "we thought it better not to trust to correspondence." The contract proposed by the railroad company involved so many important modifications, no doubt shown in mechanical drawings spread about the table, or perhaps in ad hoc sketches made in the course of the discussion, that Nasmyth and his partner declined the order.

But the trip to Nuremberg was not without enjoyment. Nasmyth was delighted with the medieval city, especially its architecture and fortifications, though he lamented the effects of growth and "progress" on it. However, progress had not affected the one place in Nuremberg that was clearly special for Nasmyth:

> I visited the house of Albert Durer, one of the greatest artists who ever lived. He was a man of universal genius—a painter, sculptor, engraver, mathematician, and engineer. He was to Germany what Leonardo da Vinci was to Italy.

Curiously, Nasmyth does not reflect upon Durer's anticipatory contribution to the rigors of mechanical drawing that were being codified in Nasmyth's time, nor does he mention the importance of Nuremberg for the most essential tool for drawing and drafting: the pencil. In all of his reminiscing about Nuremberg, James Nasmyth apparently was not aware that when he was there in 1842, the firm of A. W. Faber was operating one of the largest pencil-making operations in the world and was expanding its foreign markets. For all his attachment to sketching, which he called "the alphabet of the engineer," and thus for all his indebtedness to the common pencil as a tool of engineering thinking and communicating, Nasmyth appears to have taken no notice of it as an object of interest in itself. This can also be said to be the all-too-common fate of engineers and their method. For all that civilization owes to engineering, even engineers themselves can fail to marvel at it and its seemingly ineffable history.

ON THE BACKS OF ENVELOPES

Scientists are fond of repeating the familiar metaphor that Isaac Newton used to express his intellectual debt to Descartes and other thinkers who preceded him: "If I have seen further . . . it is by standing upon the shoulders of giants." As if to prove its own point, the metaphor was not Newton's invention; Robert Burton, the English clergyman and author of *Anatomy of Melancholy,* who died before Newton was born, had written: "A dwarf standing on the shoulders of a giant may see farther than a giant himself." And Burton acknowledged that sixteen centuries earlier, in *Civil War,* the only major Latin epic poem in which the intervention of the gods was eschewed, Lucan wrote: "Pygmies placed on the shoulders of giants see more than the giants themselves."

Scientific ideas, like literary metaphors, are seldom *sui generis,* and Newton was acknowledging precisely that when he wrote his famous line in a letter to Robert Hooke. Engineering ideas and designs also are seldom of themselves. Historically, they came out of the oral tradition of the crafts, in which nonliterary thinking abounded, and out of the pictorial catalogs of devices and structures that comprised the notebooks of early engineers. Even with the emergence of mechanical drawing conventions and, most recently, computer-aided drafting, the creative engineer still owes a debt to earlier engineers, for it is the wealth of clever artifacts, machines, and schemes they have left for us to climb upon and use that serve as the bases of ideas for the present and future. Thus the pocket calculator was first called an electronic slide rule, the pencil was named after a brush, and paper was named after the papyrus it supplanted. Engineering is the rearrangement of what is.

But all engineering starts with creative design, invariably accompanied by some kind of calculation in which the wisdom of the giants is but part of the equation. It is in this realm that the engineer's mind works wonders, with or without a pencil and paper, and sometimes at what might seem to be breakneck speed. There is a poignant anecdote from nineteenth-century bridge-building that illustrates the point. The incident took place during construction of the first bridge across the Mississippi at St. Louis, a project noted not only in the economic history of the nation for bringing together the railroads of southern Illinois and Missouri, but also in the annals of civil engineering for a bit of miscalculation that became apparent when the first of the bridge's three arches was about to be closed in 1873.

The incomplete halves of the steel arch were being supported by cables from above as the structure was formed section by section toward a meeting at the center. But when it was finally time to close the arch by lowering into the gap the last piece of steel, it would not fit, for the heat of the summer sun had so expanded the five hundred feet of metal that the gap was much narrower than had been calculated. Packing the steel with ice did not shorten it sufficiently, and waiting for cooler weather was out of the question. It was only by tightening the joints between the sections already in place, and also by tightening the cables supporting the cantilevered arch halves in order to lift the ribs, that the gap was widened enough so the last section, specially designed for the circumstances by chief engineer James Buchanan Eads, could be slipped into place. This was finally accomplished on September 17, 1873, only two days before the contract deadline, after sixty-five hours of continuous work.

A young assistant engineer, Theodore Cooper, was in charge of the nerve-racking operation, and after it was finally over he reported to Eads that he and the workmen were so sleepy that he had been afraid someone would fall from the superstructure. Since safety nets were not used in nineteenth-century construction, a fall from the bridge could be fatal. Cooper knew how easily one could fall; he himself had survived just such an experience, and in so doing had provided one of history's most dramatic examples of quick calculation. As Henry Billings recounted in his book *Bridges:*

> No one knew better than Cooper what it meant to fall from the superstructure into the Mississippi, for only a few months before, he had missed his footing on a loose plank and plummeted ninety feet to the water. It is reported that during the fall he rapidly calculated the force with which he would strike the water, while, at the same time, he had the presence of mind to roll himself into a ball. It was a deep plunge, but when he came to the surface he struck out for the rowboat that had already pushed off to pick him up. When he was hauled in, he found he was still clutching his pencil. . . .

One can imagine that Cooper was doing two things in the two and a half seconds of his fall: First, he was estimating the speed, perhaps as great as fifty miles per hour, at which he would hit the water, and the force that it might exert on him, perhaps to be better prepared psychologically and physiologically for the impact. Second, he was designing a configuration for his body that would enable it to suffer the least damage on impact. His quick calculation—which reassured him that his velocity would not be so great that he might as well die of a heart attack before striking the water—and his resulting design—to be rolled up into a ball—are two examples of what engineers do all the time in practice. As in the emergency situation in which Cooper found himself, such designs and calculations can be done quickly and almost instinctively, but generally they are accomplished with deliberation and conscious determination. In all cases, however, the selection of calculations done is made with the benefit of *experience*—the experience of having done like calculations so many times before and the experience of using the calculations of others to see the horizon and not just the deep. Whether done on the backs or the shoulders of giants, in one's head during free fall, at one's desk with the aid of pencil and paper, or verbally beside a potential client on a cliff overlooking the site for a new bridge, all such efforts are said to be "back of the envelope" designs or calculations.

The origins of the expression "back of the envelope" are obscure, but it very well may come from the fact that, at least in the days when engineers still wore suits in the field as well as in the office, an envelope was most likely tucked in one pocket or another. Perhaps the envelope contained the letter from a town official that invited the engineer to come look at the gorge that local businessmen wished to see bridged. While standing with the official and some local boosters on a promontory overlooking a chosen site, the engineer might discuss the various structural options available for a bridge, giving their pros and cons. When finally asked what his considered judgment was as to the best bridge for the site, the engineer might feel about his coat for a pencil and something to draw on, and the back of the envelope that summoned him to the town might serve as the blank leaf of an impromptu sketch pad, for a picture would be worth volumes of technical jargon.

An alternate explanation of the term's origin ties it to the sealed envelopes in which competitive bids are submitted. When selection committees are not bound to accept the lowest bid, or when the bids themselves involve competing designs of vastly different character, considerable judgment may have to be exercised in choosing among the leading proposals. Having opened the bids, singling out one among several may often have come only after some quick calculations or reality checks on the backs of the envelopes that contained the various proposals.

In Britain one hears the expression "back of the cigarette box," for unlike the cellophaned soft packs that used to be ubiquitous in America, the hard packages that British and Continental cigarette smokers were wont to carry about in their jackets could be slid open to provide not only a blank space on which to draw but also a relatively hard surface on which to do it. The cigarette box was a portable, miniature drafting board.

Regardless of whether one calculates or draws on the back of an actual envelope or on some other surface, the term "back of the envelope" has come to refer to any design that is done quickly, as a rough calculation or sketch, and usually in a very early stage, if not at the very conception, of a design. One attempt at the categorization of various kinds of engineering drawings has been made by Ken Baynes and Francis Pugh in their book *The Art of the Engineer.* According to them:

> *Designers' Drawings* . . . relate to the stage in development when the engineer is considering broad alternatives and putting forward outline schemes. They are frequently found in notebooks kept by senior engineers and are often very individual in style. It is usually this kind of drawing that people have in mind when they say that something was "designed on the back of an envelope." A characteristic of these drawings is that they leave vague those parts of the design which the designer is not concerned with at the time. They normally highlight those aspects of the design which are particularly difficult or novel.

Thus Joseph Paxton's ink-blotter sketch of the Crystal Palace and James Nasmyth's scheme-book sketches for the steam hammer could both properly, albeit metaphorically, be called "back of the envelope" sketches. And, as in both of these nineteenth-century examples, the overall concept expressed in back-of-the-envelope designs often remains remarkably unchanged through the final construction of the artifact. This should not be so surprising, for after all, the first concept serves as the basis for calculating subsequent details.

In the mid-1980s, steel was finally emerging as a popular building material in Britain, and concrete, so long dominant in the construction industry, was gradually ceasing to come first to the minds of those sketching out early structural designs. As an editorial in the London-based magazine *Civil Engineering* expressed it at the time:

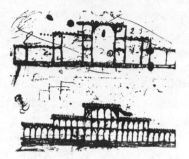

*Joseph Paxton's sketch that
resulted in the Crystal Palace*

As recently as the end of the last decade, the then President of the British Constructional Steelwork Association went on record as saying that the trouble with steel was that the first design that went on the back of the cigarette box when the consultant met with the client, was a concrete design.

Thus the design was initially figuratively and ultimately literally cast in concrete.

The first design on the back of the cigarette box or the envelope is not necessarily inferior. Indeed, for the experienced engineer, the initial design set down will be the distillation of years of direct and inherited experience and the considered, though quick, choice from among the many alternatives that may have flashed through the mind as swiftly as the alternative configurations for his body flashed through the mind of Theodore Cooper as he fell toward the Mississippi. But sometimes designs come only after years of thinking and tinkering, as in the case of an idea for an improved steamboat engine that finally came to the nineteenth-century American inventor John Stevens. According to Oliver Allen's account:

The story goes that he awoke one morning with a new scheme for the eccentrics and connecting rods and, finding

no pencil and paper handy, sketched it with his finger between the shoulder blades of his wife lying in bed next to
him. "Do you know what figure I am making?" he asked as
she awoke with a start.

"Yes, Mr. Stevens," she replied. "The figure of a fool."

Though engineers might like to think none of their schemes
foolish, whether on the back of a spouse or on the back of an envelope, careful engineers will also amplify the first sketches of
their designs with some quick and rough calculations. These are
also known as back-of-the-envelope efforts, and they are the only
kind that Cooper had time enough to make as he plummeted
toward the water.

Back-of-the-envelope calculations are meant to reveal the rea-
sonableness or ridiculousness of a design before it gets too far
beyond the first sketch. For example, one can draw on the back
of a cigarette box a design for a single-span suspension bridge
between England and France, but a quick calculation on the same
box will show that the cables, if they were to be made of any reasonable material, would have to be so heavy that they could not
even hold up their own weight, let alone that of the bridge deck.
One could also show that even if a strong enough material for the
cable could be made, the towers would have to be so tall that they
would be very unsightly and very expensive to build. Some calculations can be made so easily that engineers do not even need
a pencil and paper. That is why the designs that they discredit are
seldom even sketched in earnest, and for centuries serious designs proposed for crossing the English Channel had been either
bridges of many spans or tunnels.

The use of the computer has not eliminated the value of back-
of-the-envelope calculations. According to Mario Salvadori, who
headed his own consulting firm and who believed that "in the last
analysis all structural failures are caused by human error,"

When my engineers come to me with millions of numbers on
a high-rise, I know there is one number that tells me a lot of
things—how much the top of the building will sway in the

wind. If the computer says seven inches, and my formula, which takes thirty seconds to do on the back of an envelope, says six or eight, I say fine. If my formula says two, I know the computer results are wrong.

Computer results can be wrong for a lot of reasons, sometimes very subtle ones. Some years ago a powerful radar system was being installed to watch the skies for enemy missiles. The sophisticated system included an electronic computer that would automatically report on the trajectory of any unidentified object that might pose a threat to national security. One night, in the early stages of its use, the computer signaled an emergency condition—something was coming out of enemy territory! Fortunately, the path of the "missile" reported by the computer was so slow that no immediate retaliation was deemed necessary. In fact, the progress of the object was so deliberate that no one on duty could figure out how it kept itself up in the air. It was only when a skeptic looked out the window and saw the full moon where the computer reported the "missile" to be that a back-of-the-envelope calculation revealed that in the time it took the first radar pulse to reach and be reflected from the surface of the moon, seventy-two additional radar pulses had been emitted. Thus the computer calculated the difference in the time of return of the first pulse and the time of release of the seventy-third and declared that an intercontinental ballistic missile was located about one-seventy-third the distance between the earth and the moon. Once discovered, the bug in the program was soon corrected.

Another modern-day story involves the design of a computer-based mail system to serve the 1984 Summer Olympic Games. The success of such an innovation would naturally have depended upon whether the computer could process the mail as fast as people could put it into the system. According to Jon Bentley of Bell Laboratories, the system initially proposed by the software designers would have worked only if there were 120 seconds in each minute, for the volume of mail expected was twice the system's capacity to process it. The company designing the system was saved embarrassment only by a back-of-an-envelope recalculation by one of its managers.

Massive computer programs are now used to forecast the economy and the weather, and, as everyone knows, their predictions can be exceedingly poor. Some economists have gone on record as saying that computer forecasting is no better than that of economists using their best judgment, which is often just another way of saying, using the back of an envelope. And we all are aware that we can often predict the weather better than the meteorologist on television by just looking out the window. We also know how computers can cause the stock market to fall more precipitously than any number of traders on the floor of the New York Stock Exchange, the once-proud consumer of one million pencils annually, ever could.

But for all the individual bravado of an engineer's first design, it takes a small army of fearful calculators to bring any design to reality. The single back-of-an-envelope sketch can turn into box-cars of plans and specifications and environmental-impact statements and—in the case of projects with the implications of nuclear power plants, for example—into years of hearings, debate, and deliberation. A first design may be a triumph of elegance and individuality, but it is the drudgery and collective effort of the many designers, detailers, analysts, fabricators, constructors, inspectors, and a host of additional experts and specialists that make engineering designs work and work safely.

Distinguishing between the dwarfs and the giants may be less simple in engineering than the scientific metaphors suggest, and their total achievement is certainly greater than their individual abilities. Especially large structures such as bridges, which could cause a great loss of life if there are any lapses of attention to detail in their conception or execution, in the end should be tested not only in an engineer's mind or in a computer's processor but also in the field. Even there, though, as the test proceeds, engineers can be expected to have one hand on a pencil and another hand in a pocket or purse, reaching for an envelope or a cigarette box.

GOOD DRAWINGS AND
BAD DREAMS

A pencil sketch on the back of an envelope may capture the essence of a new engineering design, but this flash of creativity is seldom sufficient to seal a contract or build an artifact. Instead, an engineer's initial drawing will usually produce discussion, negotiation, compromise, and, throughout all this, more drawings. The first drawings beyond the first sketch are less personalized and tend to show proposals in a broader context than can be fit on—or maybe even into—a business envelope. Since they are to be the basis for communication among engineers and possibly debate about the pros and cons of a project, the drawings are produced according to generally accepted rules of drafting and conventions of the engineering profession. These rules do not, however, lighten what may be an immense emotional burden for the engineer, who knows that good drawings and careful calculations can mean the difference between triumph and embarrassment.

James Gordon, an English engineer who has written eloquently on materials and structures, both manufactured and naturally occurring, has described what can be a long process of "modifying our sketches" to get a design right, and he comments also on less private aspects of the process in an appendix to his book *Structures: Or, Why Things Don't Fall Down:*

> When all this [designing] has been done, "proper" drawings may have to be made from which the thing can be manufac-

tured. Formal engineering drawings are very necessary when components have to be made by the usual industrial procedures, but they are troublesome to make and may not be needed for simple jobs or amateur work. For anything of a commercial and potentially dangerous nature, however, it is my experience that a firm can look remarkably silly in a court of law if the only "drawing" they can produce is a sketch on the back of an envelope.

In customary modern practice, the progression of engineering drawings follows the progression of the engineering design process itself. While the word *design* is used to signify the individual act of conceptualization that puts an idea on the back of an envelope, the same word is used to signify the often long and usually collaborative process of carrying out the detailed calculations that flesh out the first sketch and thus make it possible to put specific dimensions and manufacturing instructions on the formal drawings.

There is still considerable work to do, of course, after a designer has gotten the general outline, or perhaps even several alternative outlines, of an initial concept for a bridge on paper. The early stages in the process of design, which involve more a matter of creative choice than one of analytical deduction, are accompanied by numerous sketches and drawings. According to Fritz Leonhardt, writing on the design of bridges,

If the bridge, as shown on these sketches in small scale... looks satisfactory, then the cross section can be drawn on a larger scale... in order to choose a suitable shape for the beam.... Several solutions should be drawn....

The designer should now shut himself away with these first results, meditate over them, thoroughly think over his concept and concentrate on it with closed eyes.... One then begins to draw again....

For larger bridges, one should work through one or two alternatives trying out other spans and other structural systems and make comparisons....

After several such corrective phases, a fair copy of the chosen solution will now be drawn.

While each engineer may meditate differently over a design, the arduous process of evolution described by Leonhardt is not unrepresentative. The story of the design and construction of the Golden Gate Bridge conveys the human as well as the mechanical drama of such an endeavor. In the early part of this century, San Francisco's influential city engineer Michael O'Shaughnessy often dreamed of a bridge across the Golden Gate that opened to the sea and talked freely of the idea with bridge engineers whom he met in the course of city business. One such engineer was Joseph Strauss, who headed a successful Chicago firm that made patented bascule, or counterbalanced drawbridges, of generally modest span and, at best, unremarkable appearance. When Strauss built one of his bridges in San Francisco in 1916, he and O'Shaughnessy met.

Although O'Shaughnessy had participated in many back-of-the-envelope discussions about a Golden Gate bridge, they usually involved outrageous price tags—as much as a quarter of a billion dollars—and many caveats. But Strauss's ambition to be remembered as a great bridge engineer fueled his supreme confidence that he could overcome not only any technical but also any political and financial impediments to building the world's longest bridge across one of the world's grandest vistas. And so Strauss presented an ungainly, if not a downright ugly, proposal for a combination cantilever-suspension bridge, which he said he could erect for the very attractive price of less than $25 million.

It would take years for this sketch of an ugly duckling to evolve into the portrait of a swan. In the meantime, Strauss had almost single-handedly organized the Golden Gate Bridge and Highway District, generated support for bonds to finance the project, convinced the War Department that the enormous bridge, even if it fell into the strait, would not impede the navy's ships, and answered objections from environmentalists and geologists. In the end the Great Depression made it necessary that all possible cost-cutting measures be taken, and so an all-suspension bridge without the rail lines originally planned became the choice of the engineering board that Strauss had assembled for their experience and sound judgment. The board included Leon Moisseiff, who was the designer of the Manhattan Bridge and who was chiefly responsible for the design finally adopted for the Golden Gate, and Othmar Ammann, who was chief engineer of the George Washington Bridge, then under construction between New York and New Jersey. The detailed design calculations were to be carried out by Charles Ellis, whom Strauss had lured in 1922 from his position as professor of structural and bridge engineering at the University of Illinois in Urbana.

In the early twentieth century, the education of many engineers consisted of on-the-job training. Although Ellis did not have an engineering degree—he received his A.B. in mathematics and Greek in 1900 from Wesleyan University—his mastery of mathematics enabled him to gain some early experience as a stress analyst with the American Bridge Company. He went on to do pioneer work in the field of structural engineering and wrote *Essentials in the Theory of Framed Structures,* which came to be a standard textbook on the subject. Thus, when the engineering board for the Golden Gate Bridge had finally decided on the general features of the structure, Charles Ellis was a natural choice to flesh out the design so that production drawings could be made. As he recalled later: "All the major questions requiring experience and sound judgment having been settled, there was little left to do except to design the structure. At this point, Mr. Strauss gave me some pencils and a pad of paper and told me to go to work."

Working out the details of any engineering design is no mindless chore. The size of a simple beam, for example, must be sufficient to support not only its own weight but also the load it is to carry, and with plenty of strength to spare. But the weight of a beam depends on how large it is, and how large it must be depends on how heavy it will be and what load it must carry. In the case of a suspension bridge, the size of the cables depends upon their dimensions and how heavy a roadway they must support, and determining how the shape of the cable will be changed when the heavy but flexible roadway is hung from it presents a central problem to the designer. Thus, no matter what his formal education, an engineer has to wrestle with many interrelated decisions and calculations, and as Ellis did so in Chicago, Leon Moisseiff did so independently in New York. By working within common engineering conventions they could compare notes and check each other, thus catching any errors that might creep into their reasoning or reckoning.

Staff engineers worked on subsets of calculations, but Ellis and Moisseiff checked all the results. By performing their own back-of-the-envelope calculations on the months of work of others, the experienced senior engineers could detect gross mistakes quickly. On one occasion, when a young engineer presented the results of his three months' worth of calculations on the bridge towers, Ellis wrote to Moisseiff, "You or I would not have to look at them ten seconds to know absolutely that they are very much in error." This meant, of course, that the calculations had to be redone, and such delays create great pressures on engineers, as they would on anyone. However, as John van der Zee relates in his book *The Gate*, in a situation such as that faced by Ellis and Moisseiff, it is not necessarily so time-consuming to rectify incorrect calculations or strengthen weak concepts as it is to create them from scratch:

> While pressured with deadlines, they were also able to employ, to their advantage, the blank paper syndrome. Most people, confronted with a creative or design problem, are physically incapable of putting down, on paper, a solution to it. The very emptiness of the space becomes intimidating. Yet many of these same individuals, once an idea has been conceived and committed to paper by someone else, suddenly find release. Examining a preliminary drawing, often with pen or pencil already in hand, the consciousness that was blocked before now finds itself flooded with minor improvements that can be made, ways the bare-bones idea can be adorned, corrected, embellished, redirected, saved. They are suddenly eloquent.

But Strauss, whose own graceless design was long forgotten and who was experiencing political and financial obstacles to the bridge, was not interested in eloquent engineering. As the final design work encountered delays in the tower calculations, tensions developed between Strauss and Ellis, and the latter was ordered to take an "indefinite vacation without pay." In spite of this, for a while he maintained cordial relations with Moisseiff, and Ellis even seems to have nourished the hope that he would be called back from "vacation." His interest in the bridge design never faltered, at least according to van der Zee's perhaps slightly romantic account:

> Against the weight of time that now pressed in upon him, Ellis buttressed himself with the sustaining discipline of his work. At home, alone with his drafting paper and engineer's pencils, he plunged back again into his calculations for the bridge towers, devoting weeks, then months, to the sweet, absorbing agony of revision.

Ellis never was called back. Although he had once been identified as "designing engineer" directly above the consulting engineers and directly under chief engineer Strauss on the bridge district's letterhead—which as late as 1929 carried a depiction of Strauss's ungainly design—Ellis was given no recognition in the end. Public acknowledgment of Ellis's design of the elegant towers, which are not marred by the excessive cross bracing a lesser engineer might have utilized, finally did come in a 1934 speech by Moisseiff, but it remained for van der Zee's poignant story on the occasion of the fiftieth anniversary of the Golden Gate Bridge to restore Ellis's name to prominence in the printed public record.

The human story of Charles Ellis wronged by Joseph Strauss is a sad chapter in the engineering of a great bridge, but it highlights the emotional involvement that an engineer can have with his work. The difficult engineering problem of designing the towers and designing them not only to be sound structurally but also to be pleasing aesthetically brought out the difference between Strauss the entrepreneur and Ellis the engineer. The former dreamed of being remembered as the builder of the world's greatest bridge; the latter had nightmares about the strength and safety and spirit of the structure.

The conflict between Ellis and Strauss, between humility and

hubris, has been a constant theme in engineering at least since
the twentieth century B.C., when the Code of Hammurabi required the death of the builder of any house that collapsed and
killed its owner. The tradition of the profession has evolved since
that age to one in which the engineer wants to design and build
for posterity not merely to save his own life or to immortalize
himself but also simply to design and build as well as he knows
how. To do anything else is to invite the specter of disaster, psy-
chic if not physical. The engineer's lot was described by Herbert
Hoover in his memoirs:

> If his works do not work, he is damned. That is the phantas-
> magoria that haunts his nights and dogs his days. He comes
> from the job at the end of the day resolved to calculate it
> again. He wakes in the night in a cold sweat and puts some-
> thing on paper that looks silly in the morning. All day he
> shivers at the thought of the bugs which will inevitably ap-
> pear to jolt its smooth consummation.

The metaphorical bugs that Hoover wrote about in 1952 are
the same ones that computer engineers and programmers worry
about today (notwithstanding stories that the term "bug" derives
from an actual moth found in an early computer and preserved
to this day either in Lucite in the U.S. Naval Research Laboratory or taped in a logbook kept at Harvard). In fact the term
bug was familiar to Thomas Edison, who understood that the
success of his electric light depended upon his anticipating how
it could fail. He invested considerable time and money in con-
structing a new brick building at Menlo Park, New Jersey, in
order to have a well-equipped place "to set up and test more
deliberately every point of the electric light, so as to be able to
meet and answer or obviate every objection before showing the
light to the public or offering it for sale either in this country or
in Europe." In a letter dated November 13, 1878, and addressed
to Theodore Puskas, one of the agents representing Edison
abroad, he was explicit about bugs tending to crawl into the development process:

The first step is intuition, and it comes with a burst, then difficulties arise—this thing gives out and then that—"Bugs"—as such little faults and difficulties are called—show themselves and months of intense watching, study and labor are requisite before commercial success—or failure—is certainly reached.

The anxiety of the real engineer engaged in design appears to be a universal trait, often recorded by close observers of engineers as well as by reflective engineers themselves. In 1929, at about the same time that Charles Ellis was wrestling with his assistant's erroneous tower calculations, an advertisement for cast-iron pipe that characterized the engineer as unsung hero appeared in a popular magazine. Titled "The Engineer Looks into the Future," the advertising copy began:

> When the clock hands meet at midnight he is still at work . . . dreaming over streets and structures he will never live to see. He toils behind the scenes of great civil enterprises, the unsung prophet of comforts and economics which will bless the lives of generations as yet unborn.
>
> Yet the engineer must contend with the fantasies of idealists, the rhetoric of demagogues, the lobbying of propagandists. He must check every contingency of the future against the facts and figures of today. He must bring the cool wisdom of science to every choice of methods and materials. He must properly appraise the "tremendous trifle"—the all-determining detail. Such a detail, for instance, as the kind of pipe to be used for underground gas and water mains.

While the advertisement may be self-serving for the Cast Iron Pipe Research Association, this does not invalidate the truth of its portrait. It has often been said of engineering designs that God is in the details, but no engineer wants to rely on prayer to ensure the correctness of details that are decidedly human and not divine creations. While some engineers may say their prayers before going to bed, they do not do so to leave the details to a deity.

When Robert Stephenson was designing the Britannia Bridge, he had to wrestle with conflicting advice he received about the necessity of employing chains to support the gigantic wrought-iron tubes. Whereas the theorist Eaton Hodgkinson said chains were needed, the experimentalist William Fairbairn said they were not. The question of whether to use chains, the answer to which had economic, aesthetic, and structural implications, was not one that Stephenson could leave at the office. As he admitted afterward to Thomas Gooch, onetime engineer to the Manchester & Leeds Railway,

> It was a most anxious and harassing time with me. Often at night I would lie tossing about seeking sleep in vain. The tubes filled my head. I went to bed with them and got up with them. In the grey of the morning when I looked across the square it seemed an immense distance across to the houses on the opposite side. It was nearly the same length as the span of my bridge!

Stephenson ultimately resolved his dilemma by designing towers tall enough to hold chains, should they be necessary. But other details also kept him awake nights. After the first tube had finally been floated into place, an operation of much careful planning and many anxious moments, Stephenson is reported to have remarked to Isambard Kingdom Brunel, "Now I shall go to bed."

Worrying about one's design and losing sleep over it are what many designers credit for their success. In the middle of the nineteenth century, Antonio Panizzi, who was a librarian at the British Museum, became involved with the planning and construction of a new building to house a reading room and thus became, like so many of his time, an engineer of his own creation. His rough sketch evolved into plans for the great domed reading room known to so many scholars and visitors. According to a history of the room:

> Panizzi once remarked that every shelf and peg and pivot of the new building was thought of and determined in the

wakeful hours of the night, before he communicated with anyone on the subject.

And James Gordon, writing of more recent engineering achievements, has also described the beneficial aspects of worry:

> When you have got as far as a working drawing, if the structure you propose to have made is an important one, the next thing to do, and a very right and proper thing, is to worry about it like blazes. When I was concerned with the introduction of plastic components into aircraft I used to lie awake night after night worrying about them, and I attribute the fact that none of these components ever gave trouble almost entirely to the beneficent effects of worry. It is confidence that causes accidents and worry which prevents them. So go over your sums not once or twice but again and again and again.

This is not the theoretical advice of an academic engineer but the practical advice of a veteran aircraft design engineer. And similar advice is implicit in the stories of other inventors and engineers. Oliver Evans, the American who in 1791 invented the automatic grist mill with its coordinated elevators and screw conveyors that eliminated the manual lifting and carrying of grain and flour, is reported to have put his system together in his head. Of his design Evans said, "I have in my bed viewed the whole operation with much mental anxiety." And the eighteenth-century British canal engineer James Brindley reportedly went to bed for as long as three days when faced with a difficult problem of design. When he rose, he made no plans or models, but he was ready to carry out the designs he held in his mind.

A mechanical drawing itself appeared in the dream of at least one engineer in the latter part of the nineteenth century, when J. W. C. Haldane was practicing as a civil and mechanical consulting engineer. He described the details of his "day by day career" in his *Life as an Engineer,* published in 1905. Of sleeping on problems he wrote:

When office work is not harassing through outside pressure, an engineer can sleep as well and as soundly as those gentlemen the great Caesar liked to have about him. There are occasions, however, when he is called upon to design something of a perplexing nature as rapidly as possible. No time must therefore be lost, and all the resources of past experience have to be suddenly utilized to tide him over the difficulty. For this reason he has sometimes to make the drawings himself to save time, instead of instructing assistants to do so, which thus entails very exacting employment. If, however, it is persistently continued throughout the evening, and into the small sized hours of the morning, the mind becomes somewhat strained, and upon going to bed in the early dawn the engineer may unconsciously take his work with him. Such, indeed, has been my own experience when unduly pressed.

For example, after going to bed with the intention of sleeping, I have had a rough night, or *morning* of it, having been engaged all through the state of somnolence in making calculations and drawing plans. The worst of it was, that the cantankerously unmanageable things never would come right, and, no matter what I did, every effort failed. Calculations would persist in being wrong, colours would not go on a drawing properly, and nothing I could think of was either feasible or workable. There was always something missing that had to be found, something incorrect, something unattainable, and I was glad enough to find on awaking that it had all been a dream. Such experiences have been the lot of many, and have even led to tragic results.

Thus, going to bed with a problem is no guarantee that one is going to wake up with a solution. Indeed, there have been tragic accidents as a result of some engineering detail not dealt with properly, but whether the failure was because some engineer did not lose sleep over that detail is difficult to say. It is hard to imagine that any sane engineer, whether sound sleeper or insomniac, would want a single detail of a single project to fail. Yet accidents do happen, sometimes because the project seems to be pedestrian and thus to have had all possible bugs worried out.

The collapse in early 1987 of an apartment building under construction in Bridgeport, Connecticut, occurred while the concrete floors, conveniently poured at ground level, were being hoisted into place according to a procedure known as lift-slab construction. The procedure was decades old and in recent times had been essentially error-free. Thus, not only did everyone concerned with the lifting operation apparently think it a rather routine one, but also the workmen finishing some of the details on lower floors were not even ordered out from under the tons of concrete being hoisted above them. Something did go wrong, of course, and the whole building collapsed, killing twenty-eight construction workers. In the wake of such an accident there is naturally considerable rethinking, and one consulting engineer expressed the strong opinion that contrary to the practice that had been developing, the engineers who design a structure should have a greater role at the construction site: "I am the one who has to live with the structure, and who has had the sleepless nights [while designing it]. Those worries and ideas don't necessarily get reflected in the drawings and specifications, which show the final product."

There can still be problems, of course, if engineers see their worries ignored. This is what happened when the space shuttle *Challenger* was launched in cold weather, and at least one of the engineers who advised his managers to postpone the launch was reported to have had bad dreams about something he could no longer do anything about. Much has been said of the manager-engineers taking off their engineer hats and putting on their manager hats on the evening before the fatal shuttle launch. One wonders, since they had to change hats anyway, whether things might have been different if they had put on thinking caps and worried through the night, as engineers have for centuries. But, of course, they may indeed have done just that and they may have, in their considered judgment that weighed technical, managerial, and political factors, still reached the decision to launch.

It is not only practicing engineers who seem to have learned the importance of keeping their minds under one hat while thinking about a hard problem. In his *Journal* of 1844, Ralph Waldo Emerson wrote under the heading "Otherness":

Henry Thoreau said, he knew but one secret, which was to do one thing at a time, and though he has his evenings for study, if he was in the day inventing machines for sawing his plumbago, he invents wheels all the evening and night also; and if this week he has some good reading and thoughts before him, his brain runs on that all day, whilst pencils pass through his hands.

While everyone, engineer and nonengineer alike, has experienced the feeling of being completely absorbed by whatever the mind is deeply engaged in at any given moment, it may be especially reassuring that so many engineers seem to have spent so many sleepless nights while their designs were progressing from the back of an envelope through increasingly complex and detailed calculations and drawings to the realization in an artifact upon whose safety the lives of so many may depend. If engineers do sleep, it is often with a pad and pencil nearby. They are there to record not dreams but nightmares, nightmares about collapses and explosions to be checked upon waking against the reality of a design. And it is a good thing, for otherwise there might be more tragedies than we can imagine.

FAILED PROMISES

For some time now, many of the most prominent and color-ful pages in *Mechanical Engineering* magazine have been filled by advertisements for computer software. However, there is a difference between the most recent ads and those of just a few years earlier. In 1990, for example, many software developers emphasized the reliability and ease of use of their packages, with one declaring itself the "most reliable way to take the heat, handle the pressure, and cope with the stress" while another promised to provide "trusted solutions to your design challenges."

More recent advertising copy is a bit more subdued, with fewer implied promises that the software is going to do the work of the engineer—or take the heat or responsibility. The newer message is that the buck stops with the engineer. Software packages might provide "the right tool for the job," but the engineer works the tool. A sophisticated system might be "the ultimate testing ground for your ideas," but the ideas are no longer the machine's, they are the engineer's. Options may abound in software packages, but the engineer makes a responsible choice. This is as it should be, of course, but things are not always as they should be, and that is no doubt why there have been subtle and sometimes not-so-subtle changes in technical software marketing and its implied promises.

Civil Engineering has also run software advertisements, albeit less prominent and colorful ones. Their messages, explicit or implicit, are more descriptive than promising. Nevertheless, the advertisements also contain few caveats about limitations, pitfalls, or downright errors that might be encountered in using prepackaged, often general-purpose software for a specific engineering design or analysis. The implied optimism of the software advertisements stands in sharp contrast to the concerns about the use of software that have been expressed with growing frequency in the pages of the same engineering magazines.

The American Society of Civil Engineers, publisher of *Civil Engineering* and a host of technical journals and publications full of theoretical and applied discussions of computers and their uses, has among its many committees one on "guidelines for avoiding failures caused by misuse of civil engineering software." The committee's parent organization, the Technical Council on Forensic Engineering, was the sponsor of a cautionary session on computer use at the society's 1992 annual meeting, and one presenter titled his paper, "Computers in Civil Engineering: A Time Bomb!" In simultaneous sessions at the same meeting, other equally fervid engineers were presenting computer-aided designs and analyses of structures of the future.

There is no doubt that computer-aided design, manufacturing, and engineering have provided benefits to the profession and to humankind. Engineers are attempting and completing more complex and time-consuming analyses that involve many steps (and therefore opportunities for error) and that might not have been considered practicable in slide-rule days. New hardware and software have enabled more ambitious and extensive designs to be realized, including some of the dramatic structures and ingenious machines that characterize the late twentieth century. Today's automobiles, for example, possess better crashworthiness and passenger protection because of advanced finite-element modeling, in which a complex structure such as a stylish car body is subdivided into more manageable elements, much as we might construct a gracefully curving walkway out of a large number of rectilinear bricks.

For all the achievements made possible by computers, there is growing concern in the engineering-design community that there are numerous pitfalls that can be encountered using software packages. All software begins with some fundamental assumptions that translate to fundamental limitations, but these are not always displayed prominently in advertisements. Indeed, some of the limitations of software might be equally unknown to the vendor and to the customer. Perhaps the most damaging limitation is that it can be misused or used inappropriately by an inexperienced or overconfident engineer.

The surest way to drive home the potential dangers of misplaced reliance on computer software is to recite the incontrovertible evidence of failures of structures, machines, and systems that are attributable to use or misuse of software. One such incident occurred in the North Sea in August 1991, when the concrete base of a massive Norwegian oil platform, designated Sleipner A, was being tested for leaks and mechanical operation prior to being mated with its deck.

The base of the structure consisted of two dozen circular cylindrical reinforced-concrete cells. Some of the cells were to serve as drill shafts, others as storage tanks for oil, and the remainder as ballast tanks to place and hold the platform on the sea bottom. Some of the tanks were being filled with water when the operators heard a loud bang, followed by significant vibrations and the sound of a great amount of running water. After eight minutes of trying to control the water intake, the crew abandoned the structure. About eighteen minutes after the first bang was heard, Sleipner A disappeared into the sea, and forty-five seconds later a seismic event that registered a 3 on the Richter scale was recorded in southern Norway. The event was the massive concrete base striking the sea floor.

An investigation of the structural design of Sleipner A's base found that the differential pressure on the concrete walls was too great where three cylindrical shells met and left a triangular void open to the full pressure of the sea. It is precisely in the vicinity of such complex geometry that computer-aided analysis can be so helpful, but the geometry must be modeled properly. Investigators found that "unfavorable geometrical shaping of some finite elements in the global analysis . . . in conjunction with the subsequent post-processing of the analysis results . . . led to underestimation of the shear forces at the wall supports by some 45%." (Whether or not due to the underestimation of stresses, inadequate steel reinforcement also contributed to the weakness of the design.) In short, no matter how sound and reliable the software may have been, its improper and incomplete use led to a structure that was inadequate for the loads to which it was subjected.

In its November 1991 issue, the trade journal *Offshore Engineer* reported that the errors in analysis of Sleipner A "should have been picked up by internal control procedures before construction started." The investigators also found that "not enough attention was given to the transfer of experience from previous projects." In particular, trouble with an earlier platform, Statfjord A, which suffered cracking in the same critical area, should have drawn attention to the flawed detail. (A similar neglect of prior experience occurred, of course, just before the fatal *Challenger* accident, when the importance of previous O-ring problems was minimized.)

Prior experience with complex engineering systems is not easily built into general software packages used to design advanced structures and machines. Such experience often does not exist before the software is applied, and it can be gained only by testing the products designed by the software.

A consortium headed by the Netherlands Foundation for the Coordination of Maritime Research once scheduled a series of full-scale collisions between a single- and a double-hulled ship "to test the [predictive] validity of computer modelling analysis and software." Such drastic measures are necessary because makers and users of software and computer models cannot ignore the *sine qua non* of sound engineering—broad experience with what happens in and what can go wrong in the real world.

Computer software is being used more and more to design and control large and complex systems, and in these cases it may not be the user who is to blame for accidents. Advanced aircraft such as the F-22 fighter jet employ on-board computers to keep the plane from becoming aerodynamically unstable during maneuvers. When an F-22 crashed during a test flight in 1993, according to a *New York Times* report, "a senior Air Force official suggested that the F-22's computer might not have been programmed to deal with the precise circumstances that the plane faced just before it crash-landed." What the jet was doing, however, was not unusual for a test flight. During an approach about a hundred feet above the runway, the afterburners were turned on to begin an ascent—an expected maneuver for a test pilot—when "the plane's nose began to bob up and down violently." The *Times* reported the Air Force official as saying, "It could have been a computer glitch, but we just don't know."

Those closest to questions of software safety and reliability worry a good deal about such "fly by wire" aircraft. They also worry about the growing use of computers to control everything from elevators to medical devices. The concern is not that computers should not control such things, but rather that the design and development of the software must be done with the proper checks and balances and tests to ensure reliability as much as is humanly possible.

A case study that has become increasingly familiar to software designers unfolded during the mid-1980s, when a series of accidents plagued a high-powered medical device, the Therac-25. The Therac-25 was designed by Atomic Energy of Canada Limited (AECL) to accelerate and deliver a beam of electrons at up to 25 mega-electron-volts to destroy tumors embedded in living tissue. By varying the energy level of the electrons, tumors at different depths in the body could be targeted without significantly affecting surrounding healthy tissue, because beams of higher energy delivered the maximum radiation dose deeper in the body and so could pass through the healthy parts.

Predecessors of the Therac-25 had lower peak energies and were less compact and versatile. When they were designed in the early 1970s, various protective circuits and mechanical interlocks to monitor radiation prevented patients from receiving an overdose. These earlier machines were later retrofitted with computer control, but the electrical and mechanical safety devices remained in place.

Computer control was incorporated into the Therac-25 from the outset. Some safety features that had depended on hardware were replaced with software monitoring. "This approach," according to Nancy Leveson, a leading software safety and reliabilty expert, and a student of hers, Clark Turner, "is becoming more common as companies decide that hardware interlocks and backups are not worth the expense, or they put more faith (perhaps misplaced) on software than on hardware reliability." Furthermore, when hardware is still employed, it is often controlled by software. In their extensive investigation of the Therac-25 case, Leveson and Turner recount the device's accident history, which began in Marietta, Georgia.

On June 3, 1985, at the Kennestone Regional Oncology Center, the Therac-25 was being used to provide follow-up radiation treatment for a woman who had undergone a lumpectomy. When she reported being burned, the technician told her it was impossible for the machine to do that, and she was sent home. It was only after a couple of weeks that it became evident the patient had indeed suffered a severe radiation burn. It was later estimated she received perhaps two orders of magnitude more radiation than that normally prescribed. The woman lost her breast and the use of her shoulder and arm, and she suffered great pain.

About three weeks after the incident in Georgia, another woman was undergoing Therac-25 treatment at the Ontario Cancer Foundation for a carcinoma of the cervix when she complained of a burning sensation. Within four months she died of a massive radiation overdose. Four additional cases of overdose occurred, three resulting in death. Two of these were at the Yakima Valley Memorial Hospital in Washington, in 1985 and 1987, and two at the East Texas Cancer Center, in Tyler, in March and April 1986. These latter cases are the subject of the title tale of a collection of horror stories on design, technology, and human error, *Set Phasers on Stun*, by Steven Casey.

Leveson and Turner relate the details of each of the six Therac-25 cases, including the slow and sometimes less-than-forthright process whereby the most likely cause of the overdoses was uncovered. They point out that "concluding that an accident was the result of human error is not very helpful and meaningful," and they provide an extensive analysis of the problems with the software controlling the machine.

According to Leveson and Turner, "Virtually all complex software can be made to behave in an unexpected fashion under certain conditions," and this is what appears to have happened with the Therac-25. Although they admit that to the day of their writing "some unanswered questions" remained, Leveson and Turner report in considerable detail what appears to have been a common feature in the Therac-25 accidents. The parameters for each patient's prescribed treatment were entered at the computer keyboard and displayed on the screen before the operator. There were two fundamental modes of treatment, X ray (employing the machine's full 25 mega-electron-volts) and the relatively low-power electron beam. The first was designated by typing in an "x" and the latter by an "e."

Occasionally, and evidently in at least some if not all of the accident cases, the Therac operator mistyped an "x" for an "e," but noticed the error before triggering the beam. An "edit" of the input data was performed by using the "arrow up" key to move the cursor to the incorrect entry, changing it, and then returning to the bottom of the screen, where a "beam ready" message was the operator's signal to enter an instruction to proceed, administering the radiation dose.

Unfortunately, in some cases the editing was done so quickly by the fast-typing operators that not all of the machine's functions were properly reset before the treatment was triggered. Exactly how much overdose was administered, and thus whether it was fatal, depended upon the installation, since "the number of pulses delivered in the 0.3 second that elapsed before interlock shutoff varied because the software adjusted the start-up pulse-repetition frequency to very different values on different machines."

Anomalous, eccentric, sometimes downright bizarre, and always unexpected behavior of computers and their software is what ties together the horror stories that appear in each issue of *Software Engineering Notes,* an "informal newsletter" published quarterly by the Association for Computing Machinery. Peter G. Neumann, chairman of the ACM Committee on Computers and Public Policy, is the moderator of the newsletter's regular department, "Risks to the Public in Computers and Related Systems," in which contributors pass on reports of computer errors and glitches in applications ranging from health care systems to automatic teller machines. Neumann also writes a regular column, "Inside Risks," for the magazine *Communications of the ACM,* in which he discusses some of the more generic problems with computers and software that prompt the many horror tales that get reported in newspapers, magazines, and professional journals and on electronic bulletin boards.

Unfortunately, a considerable amount of the software involved in computer-related failures and malfunctions reported in such forums is produced anonymously, packaged in a black box, and poorly documented. The Therac-25 software, for example, was designed by a programmer or programmers about whom no information was forthcoming, even during a lawsuit brought against AECL. Engineers and others who use such software might reflect upon how contrary to normal scientific and engineering practice its use can be. Responsible engineers and scientists approach new software, like a new theory, with healthy skepticism. Increasingly often, however, there is no such skepticism when the most complicated of software is employed to solve the most complex problems.

No software can ever be proven with absolute certainty to be totally error-free, and thus its design, construction, and use should be approached as cautiously as that of any major structure, machine, or system upon which human lives depend. Although the reputation and track record of software producers and their packages can be relied upon to a reasonable extent, good engineering involves checking them out. If the black box cannot be opened, a good deal of confidence in it and understanding of its operation can be inferred by testing.

The proof tests to which software is subjected should involve the simple and ordinary as well as the complex and bizarre. A lot more might be learned about a finite-element package, for example, by solving a problem whose solution is already known rather than by solving one whose answer is unknown. In the former case, something might be inferred about the limitations of the black box; in the latter, the output from the black box might bedazzle rather than enlighten. In the final analysis it is the proper attention to detail—in the human designer's mind as well as in the computer software—that causes the most complex and powerful applications to work properly.

A fundamental activity of engineering and science is making promises in the form of designs and theories, so it is not fair to discredit computer software solely on the basis that it promises to be a reliable and versatile problem-solving tool or trusted machine operator. Nevertheless, users should approach all software with prudent caution and healthy skepticism, for the history of science and engineering, including the still-young history of software engineering, is littered with failed promises.

IN CONTEXT

A recent survey of Duke University civil engineering alumni asked which of their college courses had proved to be most useful in their careers, and a significant number of respondents mentioned a sequence of history courses that was required in the 1950s. The singling out of these courses after thirty or forty years was a notable anomaly among such predictable responses as Strength of Materials, Structural Analysis, and Steel Design. Since the questionnaire did not remind the alumni of which courses they had taken, or give them a list from which to choose, one cannot help concluding that these history courses had made a clear and lasting impression on at least some students.

Alumni who listed these courses, moreover, did not identify them generically, the way they did their math, science, and engineering courses. Instead they invariably referred to them as "the history courses taught by Professor Holley." It takes nothing away from I. B. Holley, Jr., a history professor emeritus with a special interest in the history of technology who has received one of Duke's most distinguished teaching awards, to point out that the alumni have also taken classes from engineering professors of special talent and dedication. Yet none of these faculty members was mentioned so frequently in the survey, even though over the years they came in contact with many more students, many of them repeatedly in courses in their major field.

The failure of engineering teachers to be cited individually in this survey is no doubt due, at least in part, to the fact that they generally teach their courses in an impersonal manner, providing little if any historical or biographical background on the state of the art or its developers. Furthermore, the material is frequently presented in isolation from the context in which it will be employed in the real world. This kind of teaching is usually defended as necessary because of the amount of material that must be covered, often to prepare the students adequately for more of the same in subsequent courses. Also, it is argued, the analytical skills being imparted do not lend themselves to personal style or subjective interpretation.

The demands of a technical curriculum generally leave little time for telling war stories in the engineering classroom. Nevertheless, course evaluations have shown repeatedly that this is exactly the experience that students crave and remember. Students sit up and listen—as opposed to mechanically transcribing the blackboard into their notebooks—when a lecture dealing with solutions to the differential equations for a vibrating system is interrupted with stories of how the John Hancock Tower in Boston was retrofitted with tuned mass dampers to control its structural oscillations, or how wind-induced motion of the Citicorp Building in New York is mitigated by a massive block of concrete connected by springs to one of its uppermost floors. Students become engaged when discussion of the morning's news about a bridge collapse preempts the nitty-gritty of fracture mechanics that deals in abstractions for the sake of the mathematics. And students participate actively when a digression into aesthetics interrupts the objective derivation of equations relating to structural loads and deflections.

It is hardly surprising that students find engineering in its real-world context more memorable than academic exercises. Similarly, history as a recitation of successive successes, of evolutionary improvements in inventions and innovations, of things in the context only of other things, would hardly be remembered warmly by alumni. But what of history taught to make apparent the interrelationships of things and people, of structures and environment, of manufacturing and economics, of machines and war, of technology and culture, of engineering and society? What appears to have made Professor Holley's course special was the presence of a great teacher with demanding standards and down-to-earth experience—he is also a major general (retired) in the U.S. Air Force Reserve and has taught the history of technology at West Point—relating technology and other intellectual endeavors to the society and culture in which that experience is embedded.

Engineering is inextricably involved with virtually all other aspects of society, as young engineers soon learn. No engineering problem is without its cultural, social, legal, economic, environmental, aesthetic, or ethical component, and any attempt outside the classroom to approach an engineering problem as a strictly technical one will be fraught with frustration. The engineer who has been sensitized as a student to the broader nature and implications of technology begins a career with a maturity and perspective that many engineers have come to only after much grief.

From the formal beginnings of civil (as distinct from military) engineering, the relationship of the works of engineers to society has been a point of definition of the profession. In drafting a definition for the 1828 petition for a royal charter of the British Institution of Civil Engineers, Thomas Tredgold wrote that "Civil Engineering...has...changed the aspect and state of affairs in the whole world." What was not made explicit by Tredgold, most likely because it was self-evident to those seeking a royal charter, was that social, economic, and a host of other human considerations had an impact on how the sources of power in nature were directed.

The relationship between engineering and society is even more explicit in the current definition employed by the American Society of Civil Engineers, whose debt to the British Institution is evident:

> Civil engineering is the profession in which a knowledge of the mathematical and physical sciences gained by study, ex-

perience, and practice is applied with judgement to develop ways to utilize, economically, the materials and forces of nature for the progressive well-being of humanity in creating, improving and protecting the environment, in providing facilities for community living, industry and transportation, and in providing structures for the use of mankind.

Like many definitions of purpose, this one tries to cram everything about an organization of many specialized constituencies into a single sentence that will make them all happy. But such writing by committee does not detract from the fundamental message being conveyed, which is that while engineering is grounded in math and science, it serves society in very human ways. The engineering profession, in other words, does not exist for its own sake but for the sake of society. The social and cultural context in which it is embedded both motivates and constrains the very practice of the profession. I suspect that Professor Holley and his courses are remembered so long after the fact because having technology and society placed in a historical context as part of their college experience gave these engineers a leg up in their careers—as did the standard of rigor in thinking and writing that they learned.

Engineering projects have always been embedded in a social and cultural context that demanded careful attention to detail. The great pyramids, tributes to the technology of ancient Egypt, would never have been built just because the technology was in place. Indeed, it is much more likely that the pharaohs (or their chiefs of staff) challenged the engineers with a monumental problem, and the engineers developed the technology to solve it. The stories of the pyramids as case studies of design and construction under the watchful eye of a megalomaniac who dictated to the society and culture of his time and place can be valuable for understanding engineering today—just as valuable as are explications of the role of pi in shaping the pyramids, or of how their carefully crafted stones were gotten into place. Likewise, stories of cathedral-building—such as William Golding's historical novel *The Spire*, which focuses on the tension between bishop and master builder—can contribute to an understanding of the interaction between society and technology just as readily as research into the mechanical tension induced by wind can contribute to a technical understanding of masonry design.

Iron bridges and crystal palaces, among the great engineering projects of the nineteenth century, provide further examples of the complex interaction among commissioning, design, and construction. The technical component was crucial, of course, for without technological know-how and nitty-gritty engineering there would have been no hope of turning anyone's dream into reality. But it is not the now-obsolete technology that keeps the stories of Victorian engineering fresh and relevant. Rather, these accounts hold our attention because they speak of timeless forces—human nature, interpersonal dynamics, petty politics, big business, competition, and economics—that seem to be no different today from what they were then or than they were in the times of ancient pharaohs or medieval bishops. Therein lies the value of history properly taught, and when properly taught it can be as relevant as some technical courses to the practicing engineer. It was reassuring to learn recently that some of the most prestigious and accomplished engineers in America would not recoil at such an observation and may not at all have been surprised by Duke's alumni survey.

"Engineering as a Social Enterprise" was the topic of a day-long symposium sponsored by the National Academy of Engineering and held in Washington, D.C., in 1990, at the end of a year-long observance of the twenty-fifth anniversary of the group's founding. The purpose of the symposium was, according to its announcement, "to highlight the ways in which society influences technological choice, how the engineering community functions relative to the society it serves and of which it is a part, and how this relationship is changing and may evolve in the future." Since the education of engineers and nonengineers alike promised to be a central issue at the symposium, the proceedings were covered by the *Chronicle of Higher Education*.

The symposium was chaired by Walter Vincenti, professor emeritus of aeronautics and astronautics at Stanford University and author of the book *What Engineers Know and How They Know It,* whose title alone evokes questions about the nature of engineering as well as its relationship to the rest of society. As Vincenti points out, countless technological triumphs may be offered to refute any contention that there is a lack of understanding of engineering among engineers; technological efficacy, however, is not equivalent to technological savvy. The triumph of airplane design stands in contrast to the frustrations and anxieties of air travel. The successes of our interstate highway system are in contrast to the horrors of gridlock in our cities. The miracle products of chemical plants are in contrast to leaking barrels of hazardous wastes. The dream world of personal computers is in contrast to the nightmare of trying to resolve a computer billing error. Such dichotomies of modern technology seem to raise serious questions about just how much engineers really do know, especially when it comes to the impact of their work on society.

As undeniably negative consequences of technology mount, it is little wonder that Luddites are encouraged, as are lawmakers who enact legislation that introduces ill-conceived extratechnical constraints making technology itself increasingly complex. But while it may appear that engineers with little perspective have created the complexity by their initially narrow focus on the technical aspects of getting a job done, the truth of the matter is more subtle, and two-sided.

Professor Vincenti is reported to have described the academic community's view of engineering as "an exotic and incomprehensible activity that goes on 'over there.' " If such is the view of an intellectual elite, it is not surprising that engineers have been relegated to a "mainly technical role in the engine compartments of the society," as George Bugliarello—civil engineer, president emeritus of Polytechnic University, and past president of Sigma Xi—is reported to have said at the 1990 Washington symposium.

The view of engineering as exotic and incomprehensible is buttressed by the way both engineers and nonengineers are educated today. Engineering students experience a highly demanding technical curriculum, but for the most part are exposed to little if any of the history or social context of their profession. As a result, many remain unsure of themselves outside of society's "engine compartments." While all accredited engineering curricula necessarily include courses in the humanities and social sciences, engineering students, sometimes aided and abetted by their advisers, tend to view these degree requirements as obstacles to be circumvented rather than as opportunities to be seized. This story has a complement: Because nonengineering students are very unlikely to be attracted to, let alone have the prerequisites for, even the most elementary of engineering courses, our future lawyers, business executives, and policymakers, including those who are most likely to steer the ship of state, can barely speak the same language as the engineers.

There are some notable exceptions to this educational situation, of course, for there are engineering curricula that do give students a perspective on their profession; the case for the broader adoption of such curricula has been admirably argued by the civil engineer Samuel Florman in his book *The Civilized Engineer*. Furthermore, the Alfred P. Sloan Foundation has sponsored some notable attempts to address the other side of the problem by increasing "technological literacy," which includes an understanding of the engineering method among liberal arts students, and such efforts are to be applauded.

However, courses in the actual history of engineering and technology have probably the single best potential not only for giving engineers and nonengineers alike a perspective on the profession, especially with regard to its interaction with society, but also for providing the basis for a common cultural context upon which paradigms for future interactions can be based. While traditional scholarship on the history of engineering and technology, and hence the college courses based upon it, have tended to be "internalist," in the sense that they concentrated on a succession of inventions or on the description of increasingly complex technical systems with little regard for the social context into which they fit, the newer "externalist" approach looks at the bigger picture.

Among the pioneers of looking at the development of technology in context was Melvin Kranzberg, a contemporary of Holley and the primary driving force behind the founding of the Society for the History of Technology and of its journal, *Technology and Culture*. One of the leading proponents of the externalist approach to history of technology is Thomas P. Hughes, emeritus professor of the history and sociology of science at the University of Pennsylvania; his book *American Genesis* is a model of the genre. Holley's course took the externalist approach.

While Hughes's examples tend to be drawn from the history of electricity—at the National Academy symposium he spoke on the dominance of British politics over British technology in the development of that country's electrical system—timeless case studies can be drawn from every branch of engineering. The story of the Britannia Bridge, for example, which was perhaps the most watched British construction project of the late 1840s, is replete with interactions among engineering design, engineering science, local and national politics, public policy, economics, commerce, environmental matters, and more, in all their permutations and combinations. As was so much more often the case in the nineteenth century than it is now, the story of the project in so many of its technical and nontechnical aspects was admirably recorded by contemporary observers and participants. (It has also been retold and reinterpreted in the small but revealing monograph *The Britannia Bridge: The Generation and Diffusion of Technological Knowledge,* a model of interdisciplinary scholarship by the economist Nathan Rosenberg and the engineer Walter Vincenti.)

However or whenever told, the story of the Britannia Bridge shows dramatically how extratechnical considerations can force technical decisions in the design of bridges, which are often cited as the purest examples of structural engineering. In the case of the Britannia, external factors, including rights-of-way over both land and water, forced the engineer Robert Stephenson to consider alternatives to existing bridge designs and come up with the unique tubular girder that technical and economic factors were to make a dinosaur in less than a decade. The technical road down which the Britannia led British bridge design came to an abrupt end with the collapse of the Tay Bridge, while the fork in the technical road that American bridge engineers took, also in part for extratechnological reasons (recall that the premier suspension bridge builder John Roebling also manufactured and sold wire cable), led ultimately to the spectacular collapse of the Tacoma Narrows Bridge. Developing the stories of such tragedies not only provides the opportunity to teach a bit of elementary engineering (without prerequisites) but also provides a forum in which to address significant and timeless issues in engineering and society.

With the increasing literature of externalist history of engineering, Hughes wondered at the symposium about the "prolongation of innocence in engineering education." David P. Billington, professor of civil engineering at Princeton University and recipient of the prestigious Charles A. Dana Award for his innovative teaching of engineering to nonengineers, agrees with Hughes that "there is now a terrific literature available" and believes that it is time to use it. Billington's own contributions, including *The Tower and the Bridge: The New Art of Structural Engineering* and especially his more recent book *The Innovators: The Engineering Pioneers Who Made America Modern,* without sacrificing technical substance, place technology in a broad social, cultural, and aesthetic context.

Given that the motivation and resources exist for teaching engineering in a social and historical context, how is it to be done? Engineering faculties are constantly looking for places in the curriculum in which to fit more technical courses, and there seems to be a constant soul-searching about how best to include something so fundamental as design experience. Thus the symposium proposal of Rustum Roy (former director of the science, technology, and society program at Pennsylvania State University) to institute a "tithing plan," whereby at least 10 percent of an engineering student's coursework would consist of studying the history of technology and policymaking, may seem all but impossible and not likely to be taken seriously by the vast majority of engineering educators. However, it behooves us all (and in particular the Accreditation Board for Engineering and Technology, which oversees the content of engineering curricula) to take such a proposal as a starting point for considered debate on perhaps one of the most important issues facing engineering education.

That the very origins of their profession—and its roots in social goals—are generally unfamiliar to so many younger engineers today should be evidence enough that something is lacking in their education. While no informed engineer, and least of all a member of the National Academy who attended the symposium on engineering as a social enterprise, is likely to advocate diluting the technical content of the curriculum, there is clear evidence that an increasing number of accomplished and insightful engineers are concerned that the engineers of tomorrow and the rest of the world with whom they must necessarily interact will speak to each other only through an intercom between engine room and control room. For society at large that would be a most undesirable and unproductive way to proceed. But the way may be altered by a commitment to introducing well-conceived and well-taught history-of-technology courses in the curriculum. These courses can help give today's engineering students as good a preparation for entering the profession fully aware of its relationship with society as Professor Holley's students found they had forty years ago.

MEN AND WOMEN
OF PROGRESS

One of the most enduring images of nineteenth-century American inventiveness is the painting *Men of Progress,* by Christian Schussele. The artist, in what was then a not uncommon practice, executed two virtually identical oil-on-canvas versions of the work, one of which now hangs in the National Portrait Gallery in Washington, D.C., and the other of which is owned by Cooper Union in New York City. A steel engraving of the painting was made in 1862 by John Sartain, and it is the most common version seen in black-and-white reproductions. In whatever form *Men of Progress* is viewed, however, it is a striking group portrait of nineteen contemporary inventors and industrialists. The story behind the painting and the scene that its artist created provides insights into the nature of the technological landscape of the time. These in turn provide an opportunity to reflect upon how representative of inventors and engineers the group of nineteen men really was.

Christian Schussele was born in 1826 in the French province of Alsace. As a child he copied pictures he found in village churches, and by age fifteen was painting portraits from life. He began formal studies at the Strasbourg art academy, where he learned the practice of lithography, and then he moved to Paris to work with artists there. Among his early commissions was the making of drawings of the battle scenes in the gallery at Versailles, but his career was interrupted by the 1848 revolution.

Christian Schussele's Men of Progress

Schussele immigrated to the United States, settling in Philadelphia, where he knew that Cecilia Muringer, whom he had met in Paris, resided with her Alsatian lithographer father. Christian and Cecilia were soon married, and he became established as a popular and successful artist. After some of his paintings of the early 1850s were recognized by the Pennsylvania Academy of Fine Arts, he was able to devote himself entirely to his craft. Among his early works was *Franklin before the Lords in Council* (1856). It may have been this painting, whose subject was considered by many as "the godfather of all inventors," that brought Schussele to the attention of Jordan L. Mott, the commissioner of *Men of Progress*.

Mott was born in 1799, to a New York family whose ancestors had come to America in the seventeenth century. He was a sickly child who seems not to have been groomed for any particular profession, but he showed an early inclination to and talent for invention. As Mott was growing into adulthood, anthracite coal from Pennsylvania was being introduced for domestic use, but only the larger lumps were considered efficient for burning, with the smaller pieces being cast aside in waste heaps. One especially large pile of refuse coal had accumulated on the banks of the Schuylkill River at Philadelphia, and Mott acquired the lot of it for use in a new stove of his own invention, a stove that would efficiently burn small pieces of coal. In all, Mott was granted more than forty patents for coal-burning apparatus, and when iron founders could not cast stoves to Mott's specifications, he went into the business of making them himself. The J. L. Mott Iron Works, located near the employee village of Mott Haven, became very successful and earned Mott his reputation and fortune. With his background, position, and wealth firmly established in 1857, New York patron of the arts Mott commissioned Christian Schussele to paint a group portrait of men who had "altered the course of contemporary civilization." Naturally, Mott was to be among the group, and it is generally believed that he decided with whom he was to be immortalized.

Group portraits of the kind Mott commissioned were common, and by the mid-1850s Schussele had established himself as a master of the genre and thus was Mott's natural choice. The subjects of *Men of Progress* never assembled as a group, but Schussele traveled to see each of them and took about five years to produce a completed oil on canvas. Originally there were to be only eighteen inventors in the painting, but when John Ericsson's ironclad *Monitor* effectively challenged the *Merrimac* at Hampton Roads, Virginia, in March 1862, his likeness, in the form of the figure standing just to the right of the column, was added to the still-undelivered painting.

The others in the painting, from left to right, are: William Morton, a Boston dentist who invented a new system of installing false teeth and subsequently the effective use of ether anesthesia to make the process tolerable; James Bogardus, who introduced iron buildings into New York City; Samuel Colt, who conceived his revolver on a voyage to India and patented it first in France and England; Cyrus McCormick, whose automatic mowing and reaping machine was displayed at the Great Exhibition; Joseph

Saxon, who was responsible for standardizing measurement in U.S. customs houses; Charles Goodyear (seated), who succeeded in inventing a process to vulcanize rubber but failed to profit from it; Peter Cooper, who established great ironworks and designed and built the first American locomotive; Jordan Mott, whose commission assured him a central position in the painting; Joseph Henry, the experimental physicist who developed an effective electromagnet, who built the first electric motor, and who was the first and longtime secretary of the Smithsonian Institution; Eliphalet Nott, who held patents relating to steam engines and stoves and who was president of Union College for about sixty years; John Ericsson, who designed railroad locomotives as well as ironclad ships; Frederick Sickels, whose cut-off device that enabled steam engines to run more smoothly was the subject of contention for the fourteen-year duration of its patent; Samuel Morse, a model of whose telegraph is on the table and whom fellow painter Schussele may have seen as the primary focus of the group; Henry Burden, who invented machines for making horseshoes and railroad spikes; Richard Hoe, who invented the cylinder press; Erastus Bigelow, who created looms for making lace and carpet; Isaiah Jennings, whose machines made thimbles and eyelet holes; Thomas Blanchard, whose lathe could turn out gun barrels in one continuous operation; and Elias Howe, inventor of the sewing machine. Benjamin Franklin looks out over the group from the painting on the wall.

The *conversazione* at which the inventors are assembled is set in the Great Hall of the U.S. Patent Office, and thus the drawings and models, which were then required to accompany patent applications and which were put on display, are in a familiar context. Some observers have noted that Schussele used these artifacts to rank his subjects, but it might be difficult to rank the artifacts themselves. Colt's revolver sits atop what appear to be wooden boards, perhaps the remnants of a crate that held a patent model; a model of McCormick's reaper is on the floor; a pair of rubber overshoes sit beneath Goodyear's chair; Morse's telegraph is on the table; a drawing of Hoe's rotary press occupies a central position on the floor; a drawing of what appears to be one of Bigelow's looms hangs on the wall; and a model of Blanchard's lathe lies on the floor beside a model of Howe's sewing machine. There is a partially hidden model beside Burden, and books—perhaps representing science, law, or patent indexes—are scattered throughout the painting. Henry, who held no patents himself, leans heavily on a book, the medium of science, symbolizing perhaps his general opposition to the granting of letters patent. Curiously, no artifact appears to be associated with Mott himself.

The painting that now hangs in the National Portrait Gallery is signed and dated 1862. It remained with the Mott family until it became part of Henry Ford's collection. Andrew W. Mellon later acquired the work, and in 1942 he gave it, along with the National Gallery of Art itself, to the U.S. government. When President Harry Truman wanted a companion painting to *Peacemakers,* which shows President Lincoln, Generals Grant and Sherman, and Admiral David Porter aboard a steamship near the end of the Civil War, *Men of Progress* was offered. It hung in the White House from 1947 until 1965, when it was removed to the National Portrait Gallery.

The version of *Men of Progress* that belongs to Cooper Union is actually larger (about six by nine feet, compared to the Smithsonian's four by six feet) and was actually painted earlier, in 1861, with the image of Ericsson presumably added belatedly over the drapery. It was after this painting that Sartain's steel engraving was made, and this is the version most frequently reproduced. (In spite of the painting's being widely known, a recently published book on the art of British engineers includes *Men of Progress* but misidentifies it as a portrait of the competition committee for the building to house the Great Exhibition.)

Men of Progress, which had originated as a conceit of Jordan Mott, perhaps intended to decorate his comfortable living room above Fifth Avenue, soon came to hang in reproductions in much more modest surroundings all over America. It represented the young country's opportunity, greatness, and promise and had become a model for endurance, hard work, and hope. The painting, especially to those who had the benefit of John Skirving's *Key,* presented the inventor, engineer, and entrepreneur as models to be emulated by all. In this regard it was not unlike the English-man Samuel Smiles's contemporary series of books *Lives of the Engineers,* whose biographies provided examples of how success had been and could be achieved by men, even those of modest beginnings and little formal education. The Victorian mind, which was constantly being challenged with new inventions and gadgets that made life more interesting and varied, if not always more convenient, had generally come to appreciate technology and its fruits with little reservation. Those fruits could not be gotten without effort, however. As Skirving put it in his introduction to his descriptive booklet:

> The gentlemen here assembled have justly earned the title by which our print is designated—MEN OF PROGRESS; for, from their lively perception of the wants of mankind, and from their ingenuity in devising suitable appliances to provide for those needs, is derived the peculiar character and aspect of the present age as an age of progress. All honor to them—the inventors—for to them we owe the mighty triumphs of modern civilization; and the debt of gratitude due them is all the more, that theirs has been a path strewn with many thorns and few flowers; where one has succeeded in reaching the goal, many have fainted by the way. . . .
>
> Those who have achieved success have done so only by the conquest over difficulties and impediments innumerable; by hard toil, both mental and manual; by constant and untiring effort under every form of discouragement, until at length the world reaped an abundant harvest of benefits, in spite of its unwillingness.

This last paragraph applies equally, of course, to inventors and noninventors alike, to engineers and nonengineers, to men and women. Today, even with all the fervor of Skirving's words, it is difficult to overlook the homogeneity of the subjects in *Men of Progress.* Everyone in the painting is a white male who was born between 1773 and 1819 and who died between 1860 and 1895. Overwhelmingly, they were raised and flourished in the northeastern states, and we can assume that they all were part of the same old-boy network, in this case Mott's, possibly belonging to many of the same clubs and sharing many of the same ideals. What these men did for the quality of life of ordinary citizens was arguably more than that accomplished by political leaders, as many in the modest homes that displayed prints of the men may have believed. But how representative of inventors was the group? Were there no women inventors then?

Such a question must have been asked increasingly in the nineteenth century, for in 1888 the commissioner of patents compiled a list of women inventors to whom patents were issued between 1790, when the first U.S. patent was granted, and July 1, 1888. The compilation shows that during that period, out of a total of about four hundred thousand, not quite twenty-four hundred patents were issued to inventors with feminine names, or about 0.6 percent. Although some scholars have indicated that there are likely to be many errors of omission in this list, the fact remains that while women inventors were a distinct minority, they were applying for and receiving patents in the United States as early as 1809.

One indicator of the changing role of women inventors in the latter part of the nineteenth century is contained in the volumes of *Scientific American*, which gave copies of Sartain's engraving of *Men of Progress* to its subscribers. As Eugene Ferguson has reported, the weekly paper was published by Munn & Co., which also operated a patent agency, and at one time more than a fourth of all the business of the Patent Office was coming through Munn's hands. In 1850 *Scientific American* clearly saw its reader-inventor as male, and possibly a future subject for a group portrait: "Young man, whatever others have been, you can be, but not without effort—conscious, unwavering effort." By 1876, however, opportunity had apparently opened up considerably, for as one of the many booklets issued by Munn & Co. indicated, everyone could invent, "without distinction of race or color ... women or minors." All that was required to get a patent and get rich was "*to think;* not profoundly, but in a simple, easy way, which every one can do." Of course, for the services of Munn & Co., everyone without distinction also had to come up with a fee, which in the 1860s was a minimum of $25, in addition to the $35 that had to be paid to the Patent Office for a successful application.

A biographical dictionary of mechanical engineers born prior to 1861 further confirms that women were indeed involved in invention and engineering in the nineteenth century. Of 1,688 names listed in the dictionary, 6 are clearly feminine; and of the 500 singled out for biographical elaboration, at least one, Margaret Knight, was a woman. (She received many patents, including some for equipment to make paper bags.) Even assuming all of those individuals identified only by initials were men, the count of almost 0.4 percent women is in rough agreement with the Patent Office compilation. But all such sources of information can just give partial results in attempting to identify women or men engaged in invention and technology in times past. The dictionary itself, for example, reproduces *Men of Progress* as its frontispiece, but only 13 of the 19 inventors shown are even listed among the 1,688, and Mott himself is not among them.

Peter Cooper, to whom Mott appears to be listening in *Men of Progress*, was an early proponent of recognizing the contributions of women and giving them opportunities in technology. Cooper, who is said to have coinvented Jell-O with his wife, Sarah, lamented the fact that as a youth he did not get even a rudimentary education. In 1859, after he had become a successful New York businessman, he established Cooper Union "for the advancement of science and art in their application to the varied and useful purposes of life" so that "boys and girls ... who had no better opportunity" than he had had, could get an education. At night, classes were offered to both men and women in applied science and architectural drawing, and during the day women could attend the Female School of Design and take courses in art or receive "training in the new occupations of photography, telegraphy, 'type-writing' and shorthand." Classes were free, and Cooper's example is said to have influenced such other philanthropists as Andrew Carnegie, Ezra Cornell, and Matthew Vassar.

In the mid-1990s, Cooper Union's dean of engineering, Eleanor

Baum, is a prominent spokesperson for engineering education, and women comprise in excess of 15 percent of engineering students nationwide, with some programs approaching gender parity. Women are as likely as not to be at the head of their class, and they have been obviously present in such highly visible and critical technical positions as space shuttle crews and leadership of government departments and agencies. As they did in the nineteenth century, women are inventing and are applying for and receiving patents (about 5 percent of the total in 1988), and they can be expected to continue to do so in increasing numbers. It is hard to imagine a painting so exclusive as *Men of Progress* being commissioned or executed today.

SOIL MECHANICS

Children play a hand game in which a closed fist represents rock, two extended fingers scissors, and an open palm paper. At an agreed-upon signal, each of two players extends a hand in one of these configurations, and the winner is determined by the mnemonic "rock breaks scissors, scissors cut paper, paper covers rock." The same game, modified slightly, could serve to introduce the engineering enterprise of building structures on the earth, with the fist again representing rock, an extended finger structure, and an open palm soil—the relatively thin layer of "unconsolidated sediments and deposits of solid particles derived from the disintegration of rocks," otherwise known as earth or dirt. In this case, however, rock supports structure, structure displaces soil, and soil covers rock. This also suggests the sometimes surprising conditions that determine success or failure in a significant proportion of structural engineering projects.

The ideal foundation for a large structure—whether it be a suspension bridge tower, a steel skyscraper, or a concrete dam—is on sound bedrock. However, the geology underlying construction sites does not always provide such ideal conditions, and bedrock is often beyond easy reach. Furthermore, "bedrock" is sometimes inadequate as a foundation due to inherent defects. Therefore, engineers have learned to work with the hand that nature deals them. The Romans developed techniques for sinking piles into soft riverbeds, so that bridge piers rested on foundations firm and deep enough to resist the settling and scouring actions that must have led to the collapse of many an earlier bridge. The master builders of Gothic cathedrals directed the laying of broad-based foundations to provide proper footings for their edifices, lest the masonry walls crack due to uneven movement in the ground. In more recent times, engineers of tall buildings have learned to erect them over broad mats or rafts at depths that ensure sufficient buoyancy that the ground floor of the completed structure matches up with the street and does not sink over time. These and other schemes for building foundations that could not reach to bedrock have become a fundamental part of the art and science of civil engineering.

For more than eight centuries now, however, the leaning Tower of Pisa has been an effective reminder that engineers do not always build on firm foundations. Other towers and buildings, such as the Guadalupe National Shrine in Mexico City, are also noticeably tilted; some reveal the telltale signs of differential settlement through cracked walls. Some prominent structures have settled more uniformly. The Washington Monument, begun in 1848, had settled almost six inches by the time its interrupted construction was completed, in 1884, with a modified foundation. Settlement of the Palace of Fine Arts in Mexico City was more visible, for what was once its ground floor became its basement over time. Sometimes nearby construction can affect long-standing structures. This happened to Trinity Church in Boston when ground across Copley Square was excavated to build the John Hancock Tower.

Some structures are actually built of soil, which is said to be civilization's oldest building material, and failures of earth dams were endemic through the nineteenth century and have not been unknown in the twentieth. Teton Dam, a massive earth structure across the Teton River Canyon in southeastern Idaho, was built on volcanic rock that was unusually fractured. The dam failed in 1976, just six months after its completion, as its reservoir was being filled. Concrete dams have not been immune to soil problems, and the failure in 1928 of California's St. Francis Dam, some forty miles north of Los Angeles, has been attributed to a defective foundation on rock—a fault and a landslide combining to destroy the dam. The release of 12 billion gallons of water killed approximately 450 people. Explaining such failures, and obviating similar failures in future construction projects, requires an understanding of the nature and behavior of soils, especially under changing conditions of water content, consolidation, and structural loading.

The engineering science of soil mechanics, whose development was motivated by such problems, takes their solution as its challenge. It is generally agreed that the roots of soil mechanics are in the work of the eighteenth-century French civil engineer Charles Augustin Coulomb, whose studies of friction forces gave engineers a rudimentary means to determine the steepness at which a pile of sand or an earth embankment would become unstable and result in a landslide. Throughout the nineteenth century, however, few mechanical properties other than the so-called angle of repose (the steepest angle at which a pile of granular material such as sand will stay put) and bearing pressure were measured for soils, and by the end of the century Coulomb's work was discounted by working engineers. Such important soil variables as density, water content, and compressibility remained to be considered, and they would not be in any rational way until an Austrian engineer named Terzaghi found himself assigned to Turkey to lecture on construction and foundations. Fortunately that assignment, and the time for experiment and reflection that it provided, came after the engineer had already gained invaluable experience in the field. The story of this engineer's life and career, as recounted by his students and colleagues, sometimes takes on mythic proportions, something Terzaghi himself seems not to have discouraged.

Karl Terzaghi was born in Prague in 1883, when it was the capital of the Austrian province of Bohemia. Being descended from a long line of Austrian army officers, he was naturally sent to military school. But the army held no attraction for him, and he went on to attend the Technical University in Graz, where he acquired the dueling scars that marked him for life. He graduated in 1904 with a degree in mechanical engineering, a field that would never excite him; while in school he also was attracted to geology and showed promise as a writer. After further studies that concentrated on geological topics, Terzaghi began working for a Viennese engineering firm that specialized in reinforced concrete structures and hydroelectric power plants, which naturally involved the construction of dams. After three years, working mainly in the Swiss Alps, Terzaghi became restless with his assigned work, and in time he found himself in charge of the geologic and hydrographic survey for a hydroelectric power project proposed in the hinterland near the Adriatic Coast of Croatia.

Karl Terzaghi

Terzaghi afterwards secluded himself in the Alps to work on a long paper reporting on his Croatian survey. While in the Alps he heard from a friend in Petrograd of the stoppage of work on a monumental structure in that city because buildings surrounding the excavation had begun to settle and crack. Terzaghi offered to take charge of the project for the contractor, and in a short time had everything under control. Afterward he participated in the design and construction of other projects in northwestern Russia, and he became increasingly aware of the great discrepancy between the state of the art of reinforced concrete design, which was quantitative, and that of foundation design, which was not. He began to think that the problems of foundations and earthwork were due "only to gaps in our knowledge of the relationship between geological conditions and engineering consequences." The gaps could be closed, he believed, by failure analysis—that is, "by collecting and digesting case records in which each event, such as a foundation failure or the failure of an earth dam, was meticulously correlated with the geological conditions at the site."

Not surprisingly, Terzaghi was not the only person in the world struck by the paucity of understanding of soil as an engineering material. In the United States in particular, the Reclamation Service was involved in the construction of a large number of dams and irrigation systems under widely varying geologic conditions, and the director of the service agreed with Terzaghi that this was a rare opportunity to take part in a large-scale experiment that might provide much-needed insight. Thus Terzaghi spent two years in the western states observing and collecting data. But once back in Europe, he became discouraged when his attempts to make sense of it led to naught.

World War I drew Terzaghi into the Austrian army, and in 1916 he was assigned as a professor of foundations and roads in Istanbul, at the Imperial School of Engineers, which one of his former teachers was reorganizing. According to Terzaghi, "I myself felt no urge whatsoever to teach because I was too deeply pre-occupied with my own ignorance." Nevertheless, it was at this institution, now known as Istanbul Technical University, that Terzaghi was to find the opportunity to begin to reflect on his experiences and data from America and to commence some fundamental experiments on soils, employing makeshift and borrowed equipment. He soon realized the cause of his former impasse: At that time, soils were classified in strictly geological terms—coarse sand, fine sand, soft clay, stiff clay, and the like—with each such designation including soils of widely different engineering properties. Terzaghi concluded that what was needed was a means of measuring quantitatively a variety of material properties that would distinguish soils in a unique way and that would, not incidentally, enable engineers to predict by calculation such phenomena as bearing strength and settlement rate.

Such realizations were not unique to Terzaghi at the time. When the Panama Canal was being dug, landslides in the famed Culebra Cut through the continental divide had provided further dramatic and incontrovertible warnings that engineers were transgressing the limits of their ability to predict the consequences of their actions. The situation in Panama, in addition to continuing dam failures and building settlements, had led the American Society of Civil Engineers in 1913 to appoint a committee to look into such matters. This committee stressed "the importance of expressing the properties of soils by numerical values." Similar realizations came in Sweden, where railway work was accompanied by catastrophic landslides, and in Germany, where massive retaining walls were moving in unacceptable ways. But Terzaghi was to have the key theoretical insights.

After the war, Terzaghi remained in Istanbul, as a lecturer at Robert College, now known as Boğaziçi University. At Robert he taught mechanical engineering courses while pursuing his research into soils. In a reflective mood in 1918 he "wrote down in one day and on one sheet of paper" a program of experiments that he expected would take two or three years. In fact, his efforts would extend over seven years. Working alone, without access to the contemporary literature and using equipment that included cigar boxes and parts scrounged together from the college dump, he came slowly to understand the mechanics of soils. In 1920 he published a "first report" on his work, under the Americanized name of Charles Terzaghi, in *Engineering News-Record*. An accompanying editorial, "Research in Soil Mechanics," not only coined a name for the new field but also declared "the problem presented by earth as an engineering material" to be "so important that it deserves to be ranked as the outstanding research problem in civil engineering." The editorial, which also described other approaches to the problem, concluded that Terzaghi's fundamental research into determining "the mere facts of earth action" heralded "the opening of an avenue of progress which promises to lead on toward more definite knowledge of earth."

Terzaghi continued his work, and developed a model describing how water carried in microscopic voids gradually transferred the loads imposed upon it to the grain structure of clays. He first presented a mathematical model for this consolidation process in a 1923 article that received little attention. However, when he presented a paper on the topic at the First International Conference on Applied Mechanics, held in Delft, Holland, in 1924, "the response was instantaneous and enthusiastic." Terzaghi's 1925 book with the unprepossessing title *Erdbaumechanik auf bodenphysikalischer Grundlage* (Earthwork Mechanics Based on Soil Physics) led to his being offered a visiting lectureship at the Massachusetts Institute of Technology, which he accepted. Later, looking back on his seminal contributions, Terzaghi confessed that he "had only laid the foundation: the edifice remained to be created." He would also attribute his success to the fact that he "had the urge, the opportunity, and the patience in addition to the qualifications required for engineering the revolution which had already become inevitable."

With MIT as a base of operations, Terzaghi began to consult on a wide variety of earthwork and foundation problems, and he incorporated his practical experience into his teaching of soil mechanics. In 1929 he accepted a professorship at the Technical University in Vienna and thus returned to Europe. However, among the group of disciples he had left behind in America was Arthur Casagrande, a native Austrian and graduate of the Technical University in Vienna, who went from studying under Terzaghi at MIT to becoming an assistant professor in the Graduate School of Engineering at Harvard University. In the ensuing years it became increasingly clear to Casagrande and to others that the growing number of soil mechanics specialists, let alone practicing engineers, were having difficulty keeping up with the rapid developments in approaches to problems in earth and foundation engineering, and there developed a growing interest in holding a conference on the subject. The conference organized by Casagrande, which was also a means to bring Terzaghi back to the United States for the spring of 1936, proved to be a model of individual and institutional initiative.

Harvard University accepted the suggestion that it sponsor, as an official part of its tercentenary celebration, the International Conference on Soil Mechanics and Foundation Engineering, to be held in June 1936. The members in attendance were welcomed to the first official event of Harvard's celebratory year by President James B. Conant, who proceeded to review the history and development of the university. The address of the president of the conference was delivered not by the Charles Terzaghi whose articles had graced the pages of *Engineering News-Record* but by a lionized "Karl von Terzaghi, Professor at the Technische Hochschule in Vienna," and the *Proceedings* contain about a half dozen papers by "Dr. Karl v. Terzaghi."

Terzaghi returned to Vienna in the fall and attended to his expanding consulting practice. With growing unrest in Europe, however, he came back to the United States in the fall of 1938, as a visiting lecturer in soil mechanics at Harvard. That same autumn, Terzaghi was invited to lecture in Chicago, and he chose as his title, "The Dangers of Tunneling in Soft Clays Beneath Large Cities." His remarks were equally interesting to the city's Department of Subways and Traction, which had undertaken to build a subway, and the State Street Property Owners' Association, which worried about the consequences for nearby structures; both groups wished to retain Terzaghi as a consultant. He eventually chose the city, which agreed to his fee of an unprecedented $100 a day and to the stipulation that a laboratory be set up and supervised by the individual of Terzaghi's choice and under his direct supervision. Who that assistant was to be was an interesting accident of the times.

Ralph Peck, almost thirty years younger than Terzaghi, had a Doctor of Civil Engineering degree from Rensselaer Polytechnic

Institute, with an emphasis on structures and mathematics. His thesis related to the stiffness of suspension bridges, a topic inspired by the work of the prominent bridge engineer David Steinman, who was supportive of Peck's efforts. After receiving his degree in 1937, Peck attended the Detailing School of the American Bridge Company, in whose drafting room he afterward worked. When bridge work slowed down and Peck found himself laid off, however, he looked for a job wherever he could find one.

The dean at the Armour (later Illinois) Institute of Technology in Chicago told Peck there were no openings there in structures, but that if he could study either hydraulics, at the University of Iowa, or soil mechanics, at Harvard, he would have a position at Armour.

As a nondegree student at Harvard, Peck was able to join courses in the middle of the semester and took an unorthodox route to learning soil mechanics. He gained laboratory experience running consolidation tests and was sitting in on a course in statistics when Terzaghi needed someone to help him with English terminology in that field for his book in progress, *Theoretical Soil Mechanics*. Later, when Terzaghi was asked whom he wished to oversee the soil mechanics laboratory for the Chicago subway project, Casagrande's suggestion of Peck was immediately taken. The three-year project, a milestone in the development of the practical application of soil mechanics, was interrupted by the consequences of Pearl Harbor. Afterward, his work on the Chicago subway led to Peck's association not with Armour but with the University of Illinois in Urbana, where he would spend a large part of his career.

In the meantime, Terzaghi was continuing to work on his book manuscript, and Peck became involved in reviewing it. His help was acknowledged in the preface to *Theoretical Soil Mechanics*, which appeared in 1943 with the dedication, "To Harvard University, in appreciation of its liberal encouragement of the pursuit of knowledge." Even before that book was published, Peck had begun to suggest to Terzaghi that he work on a second volume, which would serve as an undergraduate textbook and would include applied aspects of soil mechanics. Terzaghi asked Peck to coauthor such a work, which would appear in 1948 as *Soil Mechanics in Engineering Practice* and become a classic text.

With his continuing teaching, publishing, and worldwide consulting, Terzaghi's reputation was secure. In 1946 he had become Professor of the Practice of Civil Engineering, a title said to have been created by Harvard especially for him. He turned down consulting jobs that did not present a fresh challenge, and he began to work increasingly in British Columbia, where the geology and foundation conditions were unusually complex. The Coast Mountains may also have evoked memories of his early work in Europe.

Among the projects that occupied Terzaghi's last years were a series of dams in British Columbia. According to Peck, one of these projects, Mission Dam, was "so complex and difficult from a geotechnical point of view that many engineers had considered it to be infeasible." Terzaghi died in 1963, and two years later, at a memorial service at the Sixth International Conference on Soil Mechanics and Foundation Engineering, in Montreal, Mission Dam was renamed Terzaghi Dam. Such an honor is rare for an engineer, but few areas of modern engineering are as inextricably linked with a single individual as soil mechanics is with Terzaghi. Indeed, the field itself stands as a memorial to his pioneering work, which so defined it.

IS TECHNOLOGY WIRED?

The rapid changes that took place in personal computers from about the early 1980s to the mid-1990s were staggering. At the beginning of this period, most home and office computers had beige complexions and monochromatic faces, and they were large in size but yet not generally hospitable enough to accommodate another thing, even one so small as a mouse. They were capable of being little more than electronic file boxes for kitchen recipes or small business inventories, inflexible word processors using floppy disks with nary the storage capacity to hold a long short story, or curious toy chests for primitive games children would laugh at today. By the end of the period, a portable personal computer could have a built-in mouse, a hard drive, a compact-disk player, a fax card, and more—and still fit inside a manila envelope. These laptop computers could be operated on the small service tray of an airplane and leave room for an evening cocktail or two. They could be plugged into telephone lines to connect to the Internet, and virtually the whole world was accessible on a flat, full-color monitor that had a faster response time than its user.

That electronic technology in the late twentieth century has been moving fast can be no surprise to anyone who has tried to keep a state-of-the-art personal computer in the home or briefcase or even to anyone who has just followed advertisements in magazines or on television. By around 1990, the vast majority of purchasers had more or less given up on waiting for the industry to reach some semblance of equilibrium, and they bought a computer when their old one broke or when it could no longer run the latest must-have software. For the layperson to hope to predict what the next year would bring was considered by and large a hopeless proposition. Computer users everywhere seemed to be resigned to acquiring and using equipment that appeared to be obsolete as soon as it was unpacked and installed on their desk or their lap.

What's with technology? Is it by nature or deliberately made a moving target that the consumer can never read or lead quite right so as to make a bull's-eye of a purchasing decision? Do engineers and marketers conspire to build planned obsolescence into their products? Or do engineers and corporate decision-makers lack the foresight to see what is just beyond the technological horizon? Do technologists lead technology, or does technology lead technologists? Is technology hard-wired in some unfathomable way that even the technologists who wire it do not fully understand? Trying to gain insight into the issues implied by such questions by looking at personal computers or any other contemporary genre of artifacts can be a frustrating quest at best, for the path to the truth is under construction and is constantly branching, curving, detouring, and dead-ending under our very feet.

Long-superseded technologies, on the other hand, can provide us with case studies as solid as fossils to help us understand the principles behind the evolution of all technology, from ancient to postmodern times. The fossilized artifacts of an earlier era, unearthed from the strata of things that are compacted in basements, attics, and old warehouses everywhere, provide facts and data that are incontrovertibly real and that are uncontroversially locatable in an evolutionary chain that has long since been stretched to its limit. Furthermore, because of the documentary evidence of contemporaneous trade catalogs, advertisements, and company archives that no longer hold any industrial secrets or consumer surprises, there are written records that enable the arrangement and interpretation of an overwhelming number of fossils of technology with the benefit of hindsight.

One very illuminating case study is that of the development of radio, as we now know it, from the concept of wireless telegraphy. The very evolution of terminology from "wireless" in the late nineteenth century to "radio" by the 1920s encapsulates the central problem associated with the technology. What the sending and receiving of audible signals through the "ether" was called not only reflected the vision and technological objectives of the inventor-entrepreneurs who dominated the technological development of the new medium of communication but also guided the planning and investment of the old and new corporations that were hoping to exploit it. Indeed, because the promise of what we now know as radio was seen from the beginning as a means of transmitting telegraph messages without the use of wires, the bulk of inventive and developmental genius was dedicated to focusing the transmitting beam to a particular receiver and maintaining the privacy of communication between two points. The potential users of the wireless telegraph service were thought to be individuals who wished their interchanges to remain confidential, and paying businesses who did not want the competition to get proprietary information free of charge. The radiating nature of the radio beam was seen as a disadvantage to be overcome through technological means, and that became the focus of much of the earliest efforts.

Telegraphy—the term relates to drawing, writing, or communicating in general over a distance—existed long before electricity. Smoke signals, drums, or even gesturing from afar were primitive forms of telegraphy. Yodeling and signing for the deaf may also be considered telegraphy, as may shouting between two people across a large room. The ancient Greeks devised alphabetical codes to transmit messages, and the development of the telescope in the seventeenth century encouraged exploitation of methods to communicate over significant distances. The semaphore came into use in late-eighteenth-century France, and Claude Chappe, its inventor, was given the title *ingénieur-télégraphe,* or "telegraph engineer." In England, the bishop-inventor George Murray devised the shutter telegraph, in which an array of shutters that could be opened and closed independently was used to transmit coded messages.

Semaphore and shutter telegraph units installed atop high elevations, often with names such as Telegraph Hill and Signal Hill, served as relay points for sending messages well beyond the horizon. Indeed, in 1837 the U.S. Congress was looking into authorizing a semaphore line between New York and New Orleans when Samuel F. B. Morse first sought government support for his telegraph system, which would operate not via visual signals but via electricity running through wires. Morse, a distinguished painter who was also interested in electrical phenomena, was following developments in electromagnetism in the late 1820s and early 1830s and was aware of such devices as Joseph Henry's new horseshoe-shaped electromagnet, which could be activated by an electrical current sent through a mile or so of wire to cause mechanical movement, such as the ringing of a bell. With his code of dots and dashes, Morse devised a system whereby an operator opening and closing a circuit by means of a device not unlike a piano key could send a coded message through intermittent current flow that would activate at the other end a sounder, whose dots and dashes a receiving operator could record. In 1843 Morse obtained enough funds from the government to string a thirty-seven-mile-long telegraph line between Washington, D.C., and Baltimore, and the line was inaugurated the following year with the famous message, "What hath God wrought."

The success of the electric telegraph was immediate and its growth enormous. However, with the proliferation of individual telegraph companies, all licensed to exploit Morse's patents, there arose complications of transferring messages from one company to the other. This soon led to the formation of mergers, including in 1851 the creation of the Mississippi Valley Printing Telegraph Company, which shortly thereafter became the Western Union Telegraph Company, under the direction of, among others, Ezra Cornell, who had strung the lines for Morse's first telegraph, using glass doorknobs as insulators, and whose fortune would found Cornell University.

Before the telegraph, the newspaper was principally a medium of opinion and a relayer and interpreter of dispatches received in the mail. In the late 1830s, to gain even a slight advantage over the competition, James Gordon Bennett, the publisher of the *New York Herald,* began the practice of sending dispatch boats to meet ships coming from Europe so the news they carried could be delivered to the newspaper office hours before the ships themselves docked. The telegraph quickly affected the way journalism was practiced, with its ability to report news virtually as it was happening. In 1848 the Associated Press was formed to pool the expenses of using the telegraph. In England, Paul Julius Reuter organized a press wire service that reached into the Continent by employing carrier pigeons where telegraph wires were still not strung. When the transatlantic cable was, after some frustrating first efforts, permanently in place in 1866, international news could be reported almost as if it were domestic.

In 1899 the America's Cup race between the yachts *Columbia* and *Shamrock* was to take place some distance off the coasts of New York's Long Island and New Jersey, and newspapers around the world were making plans to report on the race's progress. Earlier yacht contests had been covered by running a telegraph cable from a ship anchored near the course to a point on shore, with information on the status of the race transmitted as if through telegraph lines, but James Gordon Bennett, Jr., who had taken over from his father as publisher of the *New York Herald,* had a different plan. Bennett, who followed yacht races avidly, had heard of an inventor of Italian-Irish descent working in England who in the previous year had reported on a regatta by means of a new communications technique known as wireless telegraphy. Guglielmo Marconi was asked to come to New York to set up his technological system to cover the America's Cup race.

Marconi's scheme involved transmitting wireless messages from two boats anchored alongside one leg of the triangular course to receivers in downtown Manhattan and in New Jersey. From these receiving points, minute-by-minute information about the race could be transmitted overland via conventional telegraph and across the ocean via cable. Spectators who watched "Marconi bulletins" posted at the *Herald*'s offices and elsewhere throughout the city followed the race almost as if they were watching it. Marconi was an overnight celebrity, and he had dreams of launching a new technology that would transform the world.

As with all new technologies, there were shortcomings and limitations to wireless communication and plenty of competitors to point them out. For one thing, the transmissions were full of static and other extraneous noise, making them hard to hear. Furthermore, the transmissions were not private, and anyone with a receiving device tuned to the appropriate frequency could listen in. When the *Herald*, for example, had put up the money to implement Marconi's wireless idea, it did not want rival newspapers to be able to listen in on the information and possibly scoop it. Marconi's vision was to produce a wireless telegraph system that would send clear and distinct signals that only those intended to receive them could hear. This objective defined the problems that Marconi—and others—set out to solve, but in the meantime the navy adopted the new technology as a means of communicating among ships at sea. Commercial shipping was also soon equipped with wireless devices, and so there could be communication without physical or even visual contact.

Among the ships carrying wireless transmitting and receiving apparatus in 1912 was the *Titanic*. When that ship hit an iceberg late in the night of April 14, it sent out repeated SOS messages on its wireless. However, the operators on most potential receiving ships in the vicinity were no longer on duty, and the distress signals went largely unheard. In the aftermath of the sinking of the *Titanic*, federal regulation of the airwaves came swiftly and forcefully. It was no longer sufficient for ships to carry wireless sets and a single operator; at least two operators had to be aboard so that wireless signals could be monitored at all times in the event of an emergency.

The Radio Act of 1912 also placed severe restrictions on land-based amateur wireless operators, who had become a growing presence on the airwaves and were considered a nuisance because their transmissions interfered with less frivolous communications, such as those between ship and shore. By law, distress calls were given priority over other communications. All wireless operators were required to be licensed, and amateurs were effectively relegated to a passive listening role. They could only transmit in the short-wave portion of the radio spectrum at a wavelength of no more than 200 meters. Longer wavelengths were more desirable because they could be transmitted over longer distances. However, they could not be beamed in a selected direction, but rather were radiated—that is, broadcast—for receivers over a wide range to pick up and listen in on.

This broadcasting property of radio was viewed by large corporations such as American Telephone & Telegraph as a defect to be overcome if they were to develop the transmission of voice communication between a specific sender and a specific receiver that was seen to be the profitable commercial application of wireless technology. Independent inventor-entrepreneurs such as Lee De Forest, whose vacuum tube was a key element in the development of effective radio receivers, envisioned a different market, however, and pursued the development of the technology to bring news and music into the homes of people everywhere. In 1910 De Forest broadcast to the New York City area a live performance of Enrico Caruso from the Metropolitan Opera, and in 1915 De Forest was broadcasting from his factory regular nightly concerts of phonograph music. Since he also promoted his own radio apparatus during his broadcasts, he could be said to have been the first to advertise on radio.

Radio communication played an important role in World War I. In the interests of defense, amateur wireless stations were shut down by the government, and operators were encouraged to enlist in the service. They did not return to the airwaves until late 1919, but they then did so with renewed enthusiasm and a knowledge of new technology gained in the service. Improved technology also made it possible for people in households across the country to receive news and entertainment broadcasts in their living rooms or wherever they might be. Since radios at this time operated on direct current and thus required batteries, the sets could be played equally inside the home or outdoors. Although some radio sets were large and cumbersome, others were surprisingly compact and portable for the times. However, they often did require heavy batteries, and there was the problem of erecting and orientating an antenna to pick up signals effectively. In the early 1920s few of these radios came equipped with loudspeakers, and so earphones were additional ancillary equipment needed to listen in on what was being broadcast over the airwaves.

Among the noted broadcasters of the early 1920s was Frank Conrad, who worked for Westinghouse in Pittsburgh. In his spare time the amateur Conrad broadcast not only phonograph music but also live performances on a regular basis. A local department store began to advertise the availability of radio receivers costing as little as $10 that could pick up Conrad's concerts, and soon Westinghouse itself saw a whole new market for radio receivers. The company encouraged Conrad to install a more powerful transmitting station at its plant and broadcast even more frequently. The station was completed just in time to broadcast the 1920 presidential election returns, and news of this event was spread by amateurs across the country. Newspapers, on the other hand, were generally quiet about the new technology that could put them out of business.

It was amateur radio operators and word of mouth that is credited with the radio broadcasting boom of the earliest years of the 1920s. Radio sales grew at a staggering pace, going from $60 million in 1922 to more than $350 million in 1924. Americans everywhere were listening to radio, and the country was never to be the same. News was transmitted virtually as it happened, music could be listened to without having to buy records or a phonograph on which to play them, and sports events could be followed by fans everywhere as if they were in the stadium. Life on the farm and in the city was transformed, and the modern technology that was responsible was more often than not used with its bare tubes, wires, dials, and antennas proudly displayed before lone listeners or among groups that gathered to share the experience of listening in on broadcasts that were as if by magic plucked free out of thin air. The term wireless, which connoted point-to-point communication, gave way to radio as listeners reaped the pleasures and benefits of signals that radiated out in every direction from a broadcasting station.

In contrast to radio, and because of everyone's experience with it, television began deliberately as a broadcasting medium. It evolved into a wired, or cable, system first because in the early years of television there were few stations on the air and reception was poor at best for a large segment of the population who wanted to enjoy the new technology. Among the first community-antenna systems were those in small towns in the Pennsylvania mountains and that of Ed Parsons, of the coastal fishing town of Astoria, Oregon, at the mouth of the Columbia River. Parsons, who among other endeavors sold communications equipment and operated a local radio station, was urged by his wife to bring television signals to their community, and in the late 1940s he devised a system whereby he could pull in a Seattle television station and feed it via cable to customers who bought the equipment from him. By the

Radio broadcast on the occasion of the first crossing of the Delaware River Bridge, now the Benjamin Franklin Bridge, while under construction in 1924

last decade of the century, of course, cable TV was providing scores of channels to viewers even in communities with several strong signals from local stations and network affiliates, but it was facing growing competition from cableless television that received even more channels via satellite dishes that were diminishing in size even as they were growing in sophistication.

Personal computers, which increasingly began to be networked in the late 1980s and early 1990s, were first connected via cables that have become the bane of installing and moving any computer equipment. The Internet and the World Wide Web have become the routine form of electronic communication among computers, and they have provided a kind of broadcasting medium into which anyone with a computer-receiver can dip, dive, and surf. Hallways, walls, and floors of offices and homes everywhere have come to be strung with wires and cables that connect to telephone lines, ethernets, and fiber-optic cables that have come more and more to be exposed rather than installed with a sense of workmanship.

The computer network thus became a form of wired communication, and inventors and engineers everywhere were being challenged, as Marconi was with radio, to enable personal computers to communicate with each other (and sometimes privately) in a wireless way. As cellular telephones in effect fulfilled the dream of Marconi for wireless point-to-point communication, so computer users can expect developments in the not too distant future to provide them with cellular computers that capture data out of thin air in much the way that they now can listen in on, if not participate in, international radio talk shows while commuting in their automobiles. What exact form the next technologies will take is less certain than we might at first imagine, however, because tensions of the kind between wired and wireless will always pull technologists in ways of which even they cannot be fully aware. The only certain thing is that the evolution of technology, whether wired or wireless, has been, is, and always will be full of surprises, promises, and twists.

HARNESSING STEAM

The elements of the steam engine were known to the Greeks. Ctesibius of Alexandria is credited with inventing, more than twenty-one hundred years ago, the piston and cylinder, which he put to use in pumping water, and his compatriot Hero harnessed steam power to produce mechanical motion, albeit mainly to drive toys and other devices of amusement. Although the evolution of the modern steam engine is commonly reckoned from the early-eighteenth-century work of Thomas Newcomen, that is at best an oversimplification. Like all achievements in engineering, Newcomen's engine was built upon important technological developments that took place over many years in many different locations. In the case of the steam engine, events in the seventeenth century in particular laid some important groundwork and foreshadowed some ominous developments in the nineteenth and twentieth centuries.

Salomon de Caus was a French-born engineer whose activity ranged from tutoring the prince of Wales to landscaping the gardens of Heidelberg Castle. Among other things, he developed a scheme whereby steam produced in a spherical vessel could force jets of water to considerable heights in ornamental fountains. Just after midcentury, the German Otto von Guericke demonstrated the power of a vacuum created within a twenty-inch-diameter cylinder fitted with a piston: The efforts of fifty men could not prevent the piston from being sucked in. Years later, von Guericke conducted his more famous demonstration, in which teams of horses were unsuccessful in pulling apart two hollow hemispheres from which the air had been evacuated. At about the same time, a British inventor, Edward Somerset, the marquis of Worcester, devised a steam-driven water pump, known as his "water commanding engine," which reportedly could raise water as high as forty feet.

Toward the end of the seventeenth century, just before fleeing religious persecution in France, Denis Papin served briefly in Paris as a research assistant to the Dutchman Christiaan Huygens. There Papin observed some of Huygens's experiments with a gunpowder-fired engine, a forerunner of the internal-combustion concept, which Papin later pursued until he came to employ steam as a less dangerous source of energy. His engine used a fire under a closed cylinder to heat water to steam, which drove a piston. The fire was then removed to allow the steam to condense, thereby creating a vacuum that caused the piston to return to its original position. In 1698 the British mining engineer Thomas Savery patented a steam pump that could raise water as high as fifty feet. However, because many mines were much deeper than that, Savery's "miner's friend" came to be applied more for supplying water to the top floors of London buildings. Among the drawbacks of the device were the limitations of relying on atmospheric pressure and the danger of boiler explosions because of the use of steam under high pressure.

Thomas Newcomen built upon the experience developed in the seventeenth century, noting not only the features that worked but also those that were detriments to lifting water from deep mines. The scheme he devised required steam at little more than atmospheric pressure to raise a piston, water to condense the steam inside the cylinder, and atmospheric pressure to lower the piston into the evacuated cylinder and thus complete the cycle. Recognizing the superiority of Newcomen's engine, Savery allowed the use of his patent covering all water-raising devices that employed "the impellant force of fire," in exchange for partnership in the new engine.

Although Newcomen's atmospheric engine was effective in raising water from mines, to the point of being credited with revitalizing the mining industry in North-Central England, it also had its own shortcomings and limitations. Chief among these was its inefficient use of energy, for the entire cylinder was alternately heated and cooled to generate and condense the steam during each cycle. Some critics estimated the wastage to be more than 90 percent of the fuel consumed. Others used hyperbole: "It takes an iron mine to build a Newcomen engine and a coal mine to keep it going." The Scottish-born James Watt addressed this problem in the late eighteenth century by developing a condensing unit separate from the cylinder proper. By this means the cylinder could be kept at a relatively constant temperature and thus energy was not wasted in heating it anew each cycle. Watt's earliest engines consumed only about half the fuel required previously to do the same amount of work.

Late-eighteenth-century steam engines were restricted mainly to use as pumps, however, because they provided only on the downward stroke of the piston a lifting force through a flexible chain pulled upward via a rocker beam. This limitation of a purely unidirectional vertical motion did not allow the ready application of the Newcomen engine to driving factory machinery, which required a continuous rotary motion, like that produced by a water wheel. To address this failing, Watt devised the double-acting cylinder, in which steam alternately admitted and released from two sides of an enclosed piston drives it both ways. On the other side of the engine's rocking beam, Watt replaced a flexible chain with rigid linkages and gearing that converted reciprocal motion to rotary. This made the steam engine attractive to the British cotton industry, which was then the main user of automatic machinery. Thus freed of a reliance on water power, manufacturers could and did build larger factories and located them nearer to sources of raw materials and labor, such as in and around port cities.

Steam engines thus evolved from curiosities to shapers of industry and society without benefit of the engineering science of thermodynamics. Indeed, it was efforts to improve the poor

5-percent thermal efficiency of Watt's steam engine that drove the French mechanical engineer Nicolas-Léonard-Sadi Carnot's theoretical studies. Thermodynamics had little impact on steam engine design until well into the nineteenth century, but this is not to say that technological advances had to wait for thermodynamics to mature.

In developing the steam engine, Watt benefited from his partnership with Matthew Boulton, whose state-of-the-art ironworks had highly skilled workmen operating top-quality machinery. It was Boulton's entrepreneurial sense of a market for rotary power that led him to insist that Watt develop that aspect of the engine. With Boulton handling the financial end of the partnership, a research-and-development operation could be supported in which Watt pursued further improvements as well as other inventions, such as a centrifugal governor and an indicator device to record engine power output, thereby also providing a means for calculating royalties that were owed to the patentholders. To compare different engines, Watt devised the term "horsepower," which he determined to be 33,000 foot-pounds per minute, the rate, he calculated, at which a brewery horse could do work. Boulton successfully petitioned Parliament to extend Watt's original 1769 patent for a separate condenser to the end of the century. Thus, rather than struggling just to recover their investment before the original patent would have expired, in 1783, Watt and Boulton realized considerable financial gain.

Watt's machines were designed to operate at only about 7 pounds per square inch above atmospheric pressure, but steam engines could be dangerous because far greater pressures could build up in their boilers. Indeed, Papin's invention of a pressure valve and Watt's fly-ball governor were designed to keep the machines from running out of control. In time, increasing capabilities to manufacture mechanical devices with tighter tolerances and more leakproof boilers made it possible to fashion steam engines that operated under greater and greater steam pressure. No matter what safety devices were designed to limit speed and prevent overpressure in these machines, however, some operators were inclined to tie down pressure-relief valves or otherwise override safety controls.

By the middle of the nineteenth century, when the use of steam engines was widespread not only as a source of stationary power for factories but also, thanks to the likes of Robert Fulton and Richard Trevithick, as a source of motive power for steamboats and railroad locomotives, boilers were unwisely pressurized to several times atmospheric pressure, and explosions had come to be commonplace. Steamboat captains especially were prone to driving their power plants at higher-than-prudent pressures, to achieve greater speeds. As early as 1824 there were calls for restrictive federal legislation. The Franklin Institute, founded in Philadelphia that same year, came to devote much space in its journal to the subject of boiler explosions. By the early 1830s, government funds were granted to the institute for experimental apparatus to test boilers, and its report helped lead to the introduction of federal legislation, which was passed in 1838. This required independent boiler inspectors, but it specified no inspection criteria and was ultimately ineffective in reducing steamboat explosions to any significant degree. Continuing incidents finally caused Congress to create, in 1852, the Joint Regulatory Agency. This federal control did result in a diminution of deaths due to steamboat accidents, but the stationary steam engines operating in factories remained unregulated.

As has often been the case with a technology that has ceased to be novel, steam power had come to be taken somewhat for granted, and its potential dangers were largely ignored, or at least minimized. While boiler explosions were occurring in factories and on waterways everywhere, one incident in Hartford, Connecticut, in 1854, aroused more than the usual interest. On March 2, a boiler in the engine room of the Fales & Gray Car Works exploded, killing nine people immediately, with twelve more dying of injuries and fifty others being left seriously hurt. There was an inquiry, but it was to be ten years before Connecticut had a boiler-inspection law.

A more significant outcome of the Fales & Gray incident was

the initiative of several Hartford businessmen associated with the use of steam power to organize in 1857 a group known as the Polytechnic Club, which supported the rational study of the properties of steam and the causes of boiler explosions. The club did not accept prevailing superstitions and theories that explosions were caused by such things as acts of God, a demon in the boiler, or the violent recombination of the hydrogen and oxygen into which the steam had broken down. Rather, it concluded logically that explosions occurred when steam pressure exceeded the ability of the boiler to contain it. Such a situation could be remedied by the use of reliable materials, sound designs, and regular inspections to identify and correct weaknesses and deterioration in existing boilers. There was already in England the Association for the Prevention of Steam Boiler Explosions and the Boiler Insurance and Steam Power Company, the latter of which not only inspected boilers but also insured their owners and operators against losses and claims. The Polytechnic Club was well on its way to instituting similar practices in America when the Civil War caused it to disband.

In 1865, another significant boiler explosion took place, on the Mississippi steamboat *Sultana,* which was overloaded with about twenty-two hundred passengers, most of them Union soldiers recently freed after the Confederate surrender at Appomattox. The death toll, variously estimated at twelve hundred to fifteen hundred, made this the worst marine disaster in U.S. history to that time. The *Sultana* incident reawakened discussion among some Polytechnic Club members, who formed the Hartford Steam Boiler Inspection and Insurance Company and incorporated it in 1866. Before long, the company was offering services to manufacturers, including supervising the selection of materials, construction, and installation of steam boilers. It also created an engineering department to help its policyholders with boiler design. Municipal and state authorities began to accept, in lieu of their own, the company's inspections.

Boiler explosions continued to occur, however, and this blemish on advancing technology was among the concerns of a group of engineers who met in 1880 to form a new organization, the American Society of Mechanical Engineers (ASME). Other concerns of the founders of ASME were the establishment "with scientific precision" of standards for threads on nuts and bolts and of procedures for testing the strength of iron and steel.

A code of practice titled "Standard Method for Steam Boiler Trials" was formulated by ASME as early as 1884, and papers and reports began to be published that provided background material for writing a comprehensive boiler code of practice. In the meantime, no state codes were being created for stationary boilers, even though each year several hundred were exploding nationwide. Among the reasons legislators in New England, at least, gave for not acting was their belief that Hartford Steam Boiler had virtually eliminated the dangers. The situation changed in 1905, however, when a boiler explosion in a Brockton, Massachusetts, shoe factory caused 58 deaths, 117 injuries, and $250,000 in property damage.

Another explosion the following year, in Lynn, Massachusetts, made the matter of steam boilers a highly visible political issue. Massachusetts soon passed rules regulating steam boilers, and this provided impetus to ASME members to accelerate their efforts to produce a code that was promulgated by the engineering profession rather than the government. The first ASME Boiler Code, a 148-page document, carried a publication date of 1914 but was not formally approved until 1915. It represented a unique joint venture of a professional society and an insurance company.

By the time civilian nuclear power was being promoted, the ASME Boiler and Pressure Vessel Code had become well established, and it was natural to extend it to include the components of nuclear-power plants. Now the code takes up three feet of bookshelf, and ASME's monthly magazine, *Mechanical Engineering,* carries announcements of code-committee meetings and drafts of documents available for public comment, as well as queries and responses regarding the interpretation of the code.

From its inception, the creation, expansion, and maintenance of the Boiler and Pressure Vessel Code, which by 1970 attracted some fourteen thousand inquiries annually, relied largely upon the volunteer efforts of society committee members, which presented a potential for conflict of interest. In response to critics who doubted that employees of the manufacturers could serve on such committees and fairly regulate the industry to which they owed their livelihood, ASME leaders pointed out that professional engineering standards dictated that individuals could rise above selfish interests for the common good. However, a situation soon arose that cast a long shadow over such an assertion.

In 1972 ASME received a letter requesting interpretation of the code with regard to boiler feed-water indicating devices. What made the letter remarkable was that it was drafted by the chairman and vice chairman of the ASME subcommittee on heating boilers, who were, respectively, the vice president of Hartford Steam Boiler Inspection and Insurance Company and the vice president for research of McDonnell & Miller, a company that dominated the U.S. market for heating-boiler safety controls. The subcommittee's response, which was written by the chairman, was subsequently claimed to have been used by McDonnell & Miller salesmen to discredit the feed-water indicating device of a small competing manufacturer, Hydrolevel Corporation. When Hydrolevel complained to ASME, a response came from the same subcommittee on heating boilers, which by then was chaired by the McDonnell & Miller vice president.

The matter was the subject of an article in *The Wall Street Journal*, and in 1975 hearings were held before the U.S. Senate Subcommittee on Antitrust and Monopoly. Hydrolevel sued ASME, Hartford Steam Boiler, and McDonnell & Miller's parent organization, International Telephone & Telegraph Company. The latter two defendants settled out of court in 1978, but an unrepentant ASME fought the charges that it had conspired against Hydrolevel in violation of federal antitrust laws. ASME lost, and in 1979 a U.S. district court ruled that the society was guilty and assessed damages of almost $7.5 million, a large amount but somewhat less than the annual income from the sale of codes and standards publications. The appeal went all the way to the U.S. Supreme Court, which ruled against ASME in 1982.

The Hydrolevel case is an anomaly in the eighty-year history of the ASME Boiler and Pressure Vessel Code, to which so much is owed for the safety of modern steam and other power plants. This was demonstrated very dramatically in the 1979 loss-of-coolant accident in the Three Mile Island nuclear power plant, which was designed and constructed according to the code. For all the mechanical and human failures involved in that accident, it did not result in a catastrophe of the kind too often experienced in nineteenth-century factories and steamboats. Today's steam boilers, out of sight and mind in the bowels of public places and workplaces, function silently, efficiently, and without incident. Boilers and pressure vessels across the country, whether employed to provide central heating in our great institutions or hot water in our homes, are the steady and reliable workhorses of which so many of the pioneers of steam power dreamed.

THE *GREAT EASTERN*

On a visit to London some years ago, I stopped in the gift shop of the National Portrait Gallery with the intention of purchasing a supply of postcards reproducing some of the images of great Victorian engineers that hang in the museum. Among the images I most wanted to carry back across the Atlantic was one showing Isambard Kingdom Brunel—in top hat and vest, in muddy pants and boots, with cigar in mouth and hands in pockets, at once distracted-looking and relaxed, standing in a curious pose of both formality and informality, attentive to the camera yet thinking of something else, before a backdrop of massive iron chains. This famous photograph of the engineer whom Prince Philip once suggested that the profession make its hero is widely reproduced in books on early photography and on engineering, and it is so popular that the gift shop's supply was exhausted on the day I was there. In racks that contained rows upon rows of postcards of British heroes, perhaps numbering in the thousands and ranging from artists and writers to kings and queens, the space labeled to hold the Brunel was one of only a half-dozen or so empty display slots. Upon inquiring at the desk to see if someone could restock the empty slot, I was told that the card was among those most in demand and had been on order for weeks. I was certainly not the only person intrigued by the famous Victorian photograph and its subject.

Brunel stands out among the many great nineteenth-century

Isambard Kingdom Brunel
before the checking chains of
the Great Eastern

British engineers largely because of the range and scale of his
achievements, which most likely received early encouragement
from a father also deserving of note. Marc Isambard Brunel was
born in France in 1769. A royalist, he fled France's growing revolutionary turmoil in 1793 by taking a ship to New York, and soon
was engaged in a survey for a canal route between the Hudson
River and Lake Champlain. The elder Brunel's design for a Congress building in Washington was recognized with an award but
not with execution; the design was, however, later modified and
realized in a theater. After gaining American citizenship, Marc
Brunel became chief engineer of New York; as such he designed
a new cannon foundry and became involved in the defenses of
Long Island and Staten Island.

In a chance encounter with another exiled Frenchman, Brunel
became fascinated with the problem of manufacturing the blocks

needed for a ship's rigging and devised a new method that promised to save the British Admiralty enormous sums of money. By 1799 Sophia Kingdom, an Englishwoman who had been imprisoned in France, had returned to her native land, and Brunel took the opportunity to sail for England. Marc and Sophia had been corresponding across the Atlantic, and before the end of the year they were married. Brunel would eventually be knighted and die in England.

Marc and Sophia's son, Isambard Kingdom Brunel, was born in 1806 and showed an early talent for drawing, mathematics, and things mechanical. After his education at boarding school in England, he was sent to France to attend college and serve an apprenticeship to a maker of clocks and scientific instruments. At sixteen, young Isambard returned to England and began working in his father's office. Before he was twenty, the young man was engineer in charge of the project to bore a tunnel under the Thames, for which his father had devised a shield that protected the gang of workers and supported the excavation until brickwork could be installed. The work progressed slowly, but it was such a cause for wonder that sightseers were admitted to the partially completed tunnel to marvel at being under the river and to watch the workers toil. The ticket price was one shilling, and the revenue helped satisfy the impatient investors.

The tunneling progressed under increasingly poor soil conditions until, in the second year of digging, only loose gravel appeared to separate the tunnelers from the bed of the Thames. Eventually the river broke through, although the shield and completed brickwork proved to be sufficiently strong to prevent total collapse and allow everyone in the tunnel to escape the inrush of water. On inspecting the damage from a diving bell, Isambard Brunel discovered on the riverbed a massive depression leading into the tunnel. Tons of clay were dumped into the river to plug up the hole, and eventually it was sealed and the flooded tunnel pumped out. By November 1827 everything had been cleaned up, and to celebrate, the young Brunel organized an elaborate banquet under the river. In one of the tunnel arches 50 distinguished guests dined among crimson draperies, gas candelabra, and the music of a uniformed band. In the adjoining arch of the tunnel 120 miners also feasted. A painting of the event, by an anonymous artist, is as striking as the portrait of Brunel before the chains; neither image appears to have hindered the engineer's reputation, even though both could be associated with failure and frustration before success.

Work on the tunnel resumed with caution, but water broke through again barely two months after the banquet. This time several workers were drowned, and Brunel himself survived only by luck. He suffered an injured leg and then had a relapse when he returned to work too soon with undiagnosed internal injuries. He was laid up for an extended period, which gave him plenty of time to think about the failed project and what a terrible launching it was for a young engineer's reputation. Those who controlled the financial interests in the Thames Tunnel had the shield bricked up and opened the underwater chamber only as a curiosity for sightseers. Work on the tunnel was not resumed for years, and then under the supervision of Marc Brunel because his son had gone on to new projects.

After the second tunnel accident, the exhausted Isambard Brunel had been sent to recuperate in Clifton, in the heights above the inland port city of Bristol, and it was there that he made his first independent mark. He entered four designs for a suspension bridge in a competition to span the Avon River gorge, but none of his entries nor any other was judged acceptable by Thomas Telford, whose expertise had been established by his record-holding 579-foot span across the Menai Strait. Telford was invited to come up with a design of his own; his offering incorporated high Gothic towers rising from the floor of the gorge, presumably so the bridge's span would not eclipse that of the Menai. The bridge committee felt uneasy with Telford's astonishing proposal and held a second open competition, in 1830. Brunel entered a new design with towers of distinctly Egyptian influence; it was accepted in part because all others stressed their wrought-iron chains beyond the design specifications. Principally for financial reasons, however, no suspension bridge was to be completed in Brunel's lifetime. But as a memorial to him, one of his first designs was executed, with slight modifications, in the form of the Clifton Suspension Bridge, which opened in 1864 and remains in use as one of the most dramatic spans in the world.

Near the end of 1835 Brunel found time to write in his long-neglected diary that he had indeed been busy in recent years. When he tallied up the capital that would pass through his hands, the twenty-nine-year-old engineer came up with something in excess of £5 million. Suspension bridges and docks made up an almost minuscule part of the total, which was dominated by railway projects. Indeed, most prominent among Brunel's works was the Great Western Railway, which would connect London to Bristol, and eventually to points farther west.

The GWR, as the achievement is known to this day, has been called Brunel's billiard table because of its paucity of steep gradients. The railroad's construction gave the engineer the opportunity to try some daring schemes. Among the challenges Brunel faced in achieving such a flat railroad was the design of low-rise bridges such as for crossing the Thames at Maidenhead. He selected a brick arch of untried span and shallowness, which critics said would collapse as soon as the falsework was removed. It did not, and the famous bridge remains in service. He also constructed the two-mile-long Box Tunnel, then the longest railroad tunnel, which in addition to its technical achievement became famous for its orientation: The sun is said to shine through the tunnel on only one morning of the year, April 9, which just happens to be Brunel's birthday. That such an event could not occur without a similar one six months later is an astronomical fact that does not deter Brunel hagiographers from repeating the story.

One of the features Brunel incorporated into his railway, the broad gauge, did not survive. At the time he began to lay out his tracks, early railroads were by and large using a "standard" gauge of 4 feet, 8½ inches between rails, said to be a carryover from the space between the iron rails on which horses had long drawn ore carriages. Brunel argued from the start for widening the rails in order to fit the body of the railway carriage between large wheels, thus lowering the center of gravity and achieving a more steady motion with more economy. His idiosyncratic seven-foot gauge proved eventually to be a disadvantage, however, when railways became increasing linked. Well before the end of the century the broad gauge would be completely converted to the narrower standard.

Brunel's ability to prevail in his choice of gauges and other technical details of his engineering works—and to maintain the reputation that he did—are indications of the strength of his personality. This quality of the man survives in the common memory to this day, and he remains virtually a folk hero in Britain. A recent issue of *The Economist* carried a story about the frustrations facing the British government in its wanting to encourage private industry to build new roads and railways. The influential magazine suggested that folk memories of nineteenth-century disasters "produced by engineers who were wedded to particular projects" helped separate engineering from the ownership and operation of great transportation projects. This is lamentable, according to *The Economist*. The magazine article is titled "Looking for Mr Brunel," and after suggesting that today's engineers "might pick up where they left off a century ago," it begins:

Isambard Kingdom Brunel, pictured famously with a cigar in his mouth and mud on his trousers, probably spent more time talking to parliamentary committees than designing railways. Many of the great civil engineering projects of the nineteenth century were run by engineers who also had to be designers, managers and entrepreneurs all at the same time. [T]oday's engineering industry [does not] produce Brunels. Modern engineering consultants are technicians as different from nineteenth-century entrepreneurs as a cameraman is from a film-producer.

Curiously, the picture of Brunel illustrating *The Economist*'s article is cropped to separate him from the great chain backdrop of the famous photograph, thereby separating his image from one of the artifacts symbolizing his last great transportation project, one in which he was financially involved up to his ears and emotionally involved perhaps beyond reason.

The full story of this ill-fated project has its beginning in an early meeting of the board of directors of the Great Western Railway, where some reservations were expressed about the then-record length of the main line being proposed. Rather than defend his scheme, Brunel is reported to have responded with a question of his own: "Why not make it longer, and have a steamboat go from Bristol to New York and call it the *Great Western?*" According to Brunel's modern biographer, L. T. C. Rolt, the silence following this "absurd suggestion" was broken by uneasy laughter, and the meeting adjourned. One of the board members took the question seriously, however, and after a long night of discussion Brunel convinced Thomas Guppy that the scheme was possible. This was quite a feat in 1835, in light of the conventional wisdom that a steamship could not carry enough coal to make the transatlantic journey; but, after some troublesome beginnings, in 1838 the wooden-hulled S.S. *Great Western* arrived in New York Harbor in record time, thus providing an incontrovertible refutation to the conventional wisdom.

By midcentury, Brunel's reputation was secure and he was naturally involved in various ways in preparations for the Great Exhibition. Among the committees on which he sat was the one responsible for selecting a building design. After rejecting almost 250 designs submitted to it, the committee came up with its own brick monstrosity topped by a massive iron dome advocated by Brunel. The impossibility of completing such a structure in time for the opening of the exhibition, coupled with Joseph Paxton's eleventh-hour proposal for an iron-and-glass building of extraordinary simplicity and appropriateness, led Brunel and other committee members to abandon their own idea and embrace what came to be known as the Crystal Palace. Indeed, Brunel liked Paxton's concept so much that he adopted its principle for Paddington Station, the London terminus of the GWR. Today in

Paddington a statue of Brunel occupies a centrally located position, between the stairways to and from the underground station whose tiled walls are decorated with the elder Brunel's engineering drawings for his tunneling machine.

Although Brunel's second ship, the iron-hulled, screw-propellered *Great Britain,* rescued from neglect in the Falkland Islands and now restored in a dry dock in Bristol Harbor, was a fine transatlantic craft, it could not carry enough coal to sail to Australia, an increasingly important destination, and there were no ports en route where refueling could easily take place. Brunel began to consider the possibilities of a ship large enough to obviate the problem. In his notebook, amid drawings of the new Paddington Station and railway layouts, there is what Rolt describes as a "sketch of an extraordinary steamship as long as the page is wide and bristling with funnels and masts." The sketch is dated March 25, 1852, and is labeled "East India Steamship." According to Rolt, subsequent pages are filled with variations on gigantic ships, "with different masts and rigs; sometimes with both screw and paddles; sometimes with two sets of paddles." The final design for the *Great Eastern* exceeded even Brunel's largest size estimates: It would displace 32,000 tons and be 692 feet long. (The ship's hull was seen as a gigantic iron beam, not unlike the great tubes of the then recently completed Britannia Bridge.) There were numerous details to

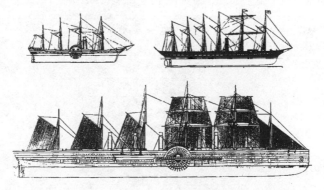

Brunel's three ships: the Great Western, *the* Great Britain, *and the* Great Eastern

worry about, of course, not least of which was where to build the ship and how to launch it.

By midcentury John Scott Russell had established himself as a preeminent engineer and naval architect. Russell ran the shipyard at Millwall, on the Isle of Dogs beside the Thames. He had become famous for his model experiments and research into the interactions of waves and ship hulls, the results of which were embodied in what came to be known as the "waveline" hull form. It was natural that the paths of Russell and Brunel should cross and that the former would be called upon to flesh out in iron what the latter had conceived in pencil. The *Great Eastern* was to be an iron ship twice as long as any then built and have a displacement an order of magnitude greater than the *Great Britain*, which had been launched almost a decade earlier.

As the ship took shape, it towered over its environs and became a tourist attraction. Reporters began to write about the growing leviathan, and one of the first major articles appeared in the *London Observer* late in 1854. Brunel's name was mentioned only once in the long article: "Mr Brunel, the Engineer of the Eastern Steam Navigation Company, approved of the project, and Mr Scott Rus-

John Scott Russell

sell undertook to carry out the design." Brunel was evidently incensed that the article was being distributed by the steamship company, and tried unsuccessfully to find out who had been the source of what he considered misinformation, appearing to give Russell more credit than Brunel wished to allow. Relations also grew strained between Brunel and Russell on matters relating to payments for work done, and by the time the ship was ready to be launched the two engineers were barely talking to each other.

The size of the *Great Eastern* and the configuration of the shipyard argued for a sideways launching of the ship, which proved to be one of Brunel's darker moments. He feared what might happen if the huge ship were to slide out of control down the rails that had been constructed beneath it, and so an elaborate launching procedure was devised that included two massive drums around which checking chains were wound to keep the ship's motion under control. Workmen were to be ready at long brake-lever arms in the event that checking was needed. It was before one of these chain-wrapped drums (and not the ship's anchor chains, as has sometimes been claimed) that Brunel's picture was taken.

The photograph is believed to be among a series that were taken before the launch was attempted on the morning of November 3, 1857. The photographer was Robert Howlett, a pioneer in the new portrait medium. His body of work is slim and little is known about Howlett, who was only twenty-eight years old when he died the following year, but his portrait of Brunel and other images captured on the launch day are remarkably sharp for the period and reveal many details of the engineer's dress and habits. The clearly visible strap across Brunel's shoulder, for example, is frequently pointed out as supporting a large cigar case that he carried about. It is variously reported to have held from fifty to more than a hundred cigars, one of which seems always to have been in Brunel's mouth. In the photograph the engineer's expression may be said to be pensive and his stare distant, which may well have characterized his mood, for he no doubt was thinking about the details of the launch and was anxious to get on with it.

First, however, the ship needed to be christened. According to the biography written by his son, "some fastidious person suggested that the name [*Great Eastern*] was objectionable, as consisting of two adjectives," and so the company directors tried to come up with something else at the last minute. Brunel was apparently making some final preparations for signaling the launch crews when he was interrupted with a list of alternative names. His oft-quoted and presumably impatient response was, "Call her *Tom Thumb* if you like." The directors evidently huddled among themselves and instructed the daughter of one of them, a Miss Hope, to christen the ship *Leviathan*. That name never stuck, however, but *Great Eastern* did.

It was finally time to get to the real business of the day, and Brunel alone was to give the signal to launch. Fastenings would be let go and the chains made slack. The hauling chains that extended around sheaves mounted on barges moored in the river would be tensed by the winches onshore. Additional winches were on the barges, and hydraulic rams were ready at bow and stern to be called into service should the ship stick on the rails. Brunel was to have complete charge of the operation, and strict silence was to be observed so that signals could be heard. Unfortunately, the directors had authorized admission tickets to be sold to the public, and the launch site was a mass of confusion.

As soon as some semblance of order was obtained, the signal to begin was given and the ship began to move, first at the bow and then very quickly at the stern. When the slack on the chains was spent, the stern drum began to revolve—so quickly that the men holding on to its handles were thrown about and one was killed.

The ship had moved only about 4 feet down the ways when the launch came to a halt. How to move the mass of iron the rest of the 240 feet along the one-in-twelve slope became a matter of much speculation. There were many kibitzers hovering about the launch site, and among the suggestions was one to have 500 troops march in step around the deck to induce vibrations that would keep the ship moving once it was started. Brunel ignored such proposals and ordered more hydraulic rams.

The ebb and flow of the tides dictated when another launch could be attempted, and the next opportunity came on November 19, when the ship was found to be stuck. New tackle and additional rams were ordered, but several subsequent attempts moved the ship only 8 or 10 feet at a time down the launchways. It was not until January 30, 1858, that the ship could finally float freely. The *Great Eastern* had been launched, but at a total construction price of £750,000 she was already a full 100 percent over budget and yet to be fitted out. Brunel's reputation had suffered considerably, his personal fortune was decimated, and his health was compromised.

Though the portrait of Brunel before the chains may be his most celebrated photograph, another that Howlett took that morning was equally interesting to George Emmerson, whose revisionist biography of Scott Russell was published twenty years after Rolt's of Brunel. Emmerson opens his story by discussing the only one of Howlett's photographs in which Russell is believed to appear. In the photo, Brunel stands between an assistant and associate on the launch platform "in characteristic pose, his legs astride in the manner of Holbein's Henry VIII, thumbs in waistcoat pockets and inevitable cigar projecting defiantly from the side of his mouth." The description is Emmerson's, and his position on Brunel versus Russell becomes clearer as he goes on to describe another figure on the platform:

> Farther over to Brunel's right, with a suggestion of independence from the trio just named, stands the remaining figure, handsome, of medium height, well built and with a proprietary air. This latter person is John Scott Russell, Brunel's engineering partner in the project, the naval architect and contractor of the Great Ship.
>
> Now at the age of forty-nine, his name was known wherever ships were built and navies assembled ... and the most influential period of his life lay ahead. Time was to reveal him as zealous educator, an idealistic social reformer and

*Scott Russell, Brunel, and others at the first attempt
to launch the* Great Eastern

even a would-be peacemaker between nations, as well as the
undisputed and respected leader of his profession. Yet there
fell a shadow on his expectations and deserts, a strange
antipathy of fortune hinted at by his division from the others
in that historic photograph. This division was exploited by
L. T. C. Rolt in his much acclaimed biography of Brunel....
Yet, because the details and facts of Russell's fascinating life
have never been investigated, this unjust calumny has since
been trustingly repeated and even expanded by innumerable
journalists and writers on nineteenth-century engineering.

Biographers, like engineers, become attached to their designs,
and once they choose between Design A and Design B there is

often little looking back at the rejected one. The good features of
the chosen design can be praised; its faults can be excused in contrast to the imagined greater ones of the spurned alternative. The
preeminent historian of British engineering, R. A. Buchanan, has
come to Rolt's defense in his depiction of Brunel, but Emmerson
has held his ground in defense of Russell in a spirited exchange
in the pages of *Technology and Culture.* Certainly neither Brunel
nor Russell was flawless, and their respective egos were no doubt
tested to the limits in the great project of building the world's
largest ship. The crucial decisions that, once made, could scarcely
be reversed, the inevitable setbacks that must necessarily accompany an endeavor of such magnitude, would try the patience and
test the mettle of anyone.

The tension between Brunel and Russell may have been of a
kind inevitable between collaborators. While there is little doubt
that Brunel deserves undisputed claim to the conceptual design
of the *Great Eastern,* he could no more have executed it alone
than could an architect single-handedly erect a skyscraper. The
problem seems to be endemic to large building projects, in which
so much creative structural detail must supplement the broad
conceptual lines of a bold drawing. Engineering spans the spectrum of creative problem-solving from concept to construction,
and there is room for many kinds of talents along the continuum.
A conceptual design is a *sine qua non,* of course, but then so are
the detailed decisions of execution. The *Great Eastern* might have
remained a sketch in Brunel's notebook had there not been a
Scott Russell to flesh it out in iron. Yet all of Russell's knowledge
of waves and hull lines might never have even so much as
brought to him in a dream the vision of a ship the scale of the
Great Eastern.

One vision that eventually did come to Russell was not a great
ship but a great book. In the wake of the *Great Eastern* experience
Russell completed his magnum opus, *The Modern System of Naval
Architecture.* The three-volume work was published in London in
1865, and Russell's masterpiece continues to have among a segment of engineers today a reputation that precedes the work it-

self. In a recent manuscript of an article in which I discussed the limitations to size that ocean wave action was believed to place on ships in the nineteenth century, I reproduced an illustration of the hogging and sagging that wave crests and troughs induced in ship hulls and attributed the oft-reproduced drawings to William Fairbairn, who founded the shipyard at Millwall and whose pioneering efforts in iron-hulled ships were at least as significant as those of both Russell and Brunel. My manuscript was returned to me with generally favorable reviews, but one of the referees questioned my attribution of the ship illustrations to Fairbairn. The anonymous peer reviewer thought that surely I must be mistaken and that the illustrations were from Russell, though no specific citation was given.

Ships on waves

I had reproduced the illustrations from Fairbairn's contemporaneous *Treatise on Iron Shipbuilding*, but to see in fact how and in what context they appeared in Russell's work, I looked for it in the library. Not finding it in Duke's collection, I asked our engineering librarian if he could locate a library that would send the volumes on interlibrary loan, and he did. Princeton was willing to send the Russell to Duke but seemed unusually concerned about how the volumes were to be used and when they would be returned, concerns that my librarian attributed perhaps to the age and condition of the volumes.

By the time this source of the Russell volumes was identified, I had gotten tied up with deadlines relating to another project, and so I suggested postponing the shipment of the books until I knew I would have some time to spend with them. The following summer my librarian gave the necessary assurances to his colleagues in Princeton, who then filled the request. In these times when rare books are cannibalized and individuals and institutions often seem so competitive, suspicious, and litigious (though perhaps they are in fact no more so than in the times of Brunel and Russell), the interlibrary loan system and the librarians who communicate through it never cease to provide reassurances that cooperation, trust, and handshakes can still guide interpersonal and interinstitutional relationships. The Scott Russell books proved to be a prime example.

The arrangements having been made to have *The Modern System of Naval Architecture* sent to Duke, I thought little more about it. As with so many of the interlibrary loans I had used over the years, I fully expected to find the Russell books in my mailbox as soon as they arrived in the engineering library. I had my mind on other things the day I walked into Perkins Library, where I often worked in the mornings and where all interlibrary loan books are received and logged in before being distributed to the branch libraries and patrons. By accident I met one of the interlibrary loan staff members, who told me that my books from Princeton had arrived and that I might want to use them in Perkins. I took this as small talk and said I'd prefer they be sent to Engineering, as usual, but he said I should really come to his office and look at them first. Even when he suggested on the way that I could borrow a book cart to move them to my carrel, I was not prepared for what I saw when I first laid eyes on the actual books. All through our discussions relating to locating and arranging to borrow the *Modern System,* neither my librarian nor I paid any attention to the size of the volumes, even though that information appears on library records no less visibly than do catalog numbers and bibliographical data. But for all our fundamental dealings with artifacts and their scale, even we engineers can sometimes overlook the effects of size on the performance and use of made things.

The Modern System of Naval Architecture was published in a very large format (twenty-eight by twenty by as many as two inches thick, with each volume having a weight to match its bulk), which made the work as unwieldy as Audubon's *Birds of America.* Indeed, the many large and exquisite mechanical and structural drawings that the Russell contains would appear to make it as likely a candidate as the Audubon to be displayed in the locked glass cases in library foyers, were it not for the bibliophilic and scholarly prejudices that value the images of natural history and philosophy over those of the mechanical and structural engineering of the Victorian era. Yet the engineering drawings that fold out to many multiples of their volume's width capture the beauty of the imagined and made world as surely as the birds of Audubon appear to capture nature. For all their associations with commerce and war, the ships of Russell convey a sense of art, calm, and reflectiveness I have never found in Audubon's prints, drawn not from life but from the carcasses the naturalist struck in poses by employing a wire skeleton after he downed his subjects with his gun.

When I recovered from the shock not only over the size and beauty of the Russell volumes but also over the fact that Princeton had entrusted them to the U.S. mails and to me, I wheeled them onto the elevator to take upstairs. My carrel proved not to have enough desk space on which to lay the books open, and so I took them to a large library table. Even here, with the books opened flat, I found them difficult to use. The type was large enough, but the words were strung out in lines that, even allowing for ample margins, were excruciatingly long. My whole head had to rotate back and forth lest my eyes tire in their sockets, and I had to hold my place on the left of the page lest I lose it in reading from one line to the next. I could not sit and comfortably read from top to bottom of each page and found myself standing each time I turned over a leaf. The book was not at all what today would be called user-friendly, and I reflected on this as I read in the Preface:

We the passing generation have had to grope our way out of the dark slowly and painfully, with trial and error. But what has to be pardoned to us can no longer be pardoned to our successors, to whom we bequeath the costly knowledge and painful experiences that have cost us so dear, but which we have gladly earned, and now painstakingly contribute for their instruction, and the advancement of their future.

Book design no less than ship design or any other kind of design is a matter of looking backward and forward at the same time. The grandeur of a volume's illustrations may so dominate its design that portability and readability become subsumed in the dominant objective. Russell's words march slowly across the twenty-inch pages, and for all their wisdom and experience, they appear to be read by few engineers or any other readers today. I had an objective, however, to find the illustrations that had led me to these tomes, and so I turned every leaf of every volume and opened every folding plate. But my search was in vain, for nowhere did I find the illustrations that I already knew to be Fairbairn's and that I was looking for now only to see how closely Russell might have copied them. I discovered no such derivative scholarship, but I did come away from the *Modern System* with a renewed respect for the man Emmerson called "a great Victorian engineer and naval architect," even in his discussion of the "blunder made in launching the *Great Eastern*." This brings my thoughts back to Brunel.

According to Russell, the blunder lay in the "experiment of launching her in iron ways." Brunel had commissioned William Froude, who was later to gain fame for his testing of model ships, to conduct experiments on the friction between iron surfaces. It was partially on the basis of those experiments that Brunel devised the details of the launching scheme for the *Great Eastern* that proved to be so tragic an embarrassment in his last years. He conceived a ship of unprecedented size and then conceived a launch scheme of an unprecedented nature and stubbornly stuck to it. His experience occurred, and his real experiments thus took place, in the building and launching of the ship itself. There were no precedents of equal magnitude from which to learn, which made it an engineering problem of the most challenging kind. But that is not to say that such projects are doomed to failure, for they are not. Indeed, the very uniqueness of such situations has always driven engineers to proceed with especial caution, and so Brunel's insistence of a controlled launch was done precisely because he knew he had no experience of what surprises were in store for him and his great ship should it slide uncontrollably into the water. In the end he had to allow the experiment of the *Great Eastern*'s launch to be played out in public over a period of three months, but when the leviathan did finally float, he was vindicated. But within two years the sickly Brunel died at age fifty-three.

The *Great Eastern* went on to lay the Atlantic cable, among other noble service, and the ship was pictured as the crowning achievement in an obituary portrait of Brunel that appeared in the *Illustrated London News*. For all its setbacks during its construction, launch, and initial sailing, the great ship captured the imagination of engineers and laypersons alike. It is not remembered as Brunel's folly but as his final challenge, and he is recalled as a great engineer precisely because he dared to focus on projects that to some appeared to be on the edge of possibility. In their successful, albeit sometimes tortuous execution, Brunel gave to the Victorian age a pride of accomplishment that lives on in Britain today, in spite of the debates among scholars. The great engineer Brunel achieved in iron what great artists achieved in oils, great poets in words, and great statesmen in deeds. Portraits of them all now hang in the National Portrait Gallery, and postcards of their images are arrayed in its gift shop.

Of all the images, Brunel's before the chains remains one of the most striking and most popular, perhaps because the mythic scale of his efforts unites art and poetry and politics with the fundamental human urge to build bigger and grander designs than have ever been tried before. For that kind of vision, the likes of Brunel would seem always to have a special place in the hearts of those of us whose projects are of a more modest scale. But no matter how grand their designs, even the Brunels appear to be subject to the very human qualities of jealousy and self-doubt and fear that their greatest experiments will not succeed. Such manifestations of humanity make the likes of Brunel—and Scott Russell—all the more believable and therefore all the more likable as heroes, however flawed.

DRIVEN BY ECONOMICS

By the first morning out of Southampton on the *Queen Elizabeth 2,* all eighteen-hundred-odd passengers were irrevocably committed to a greatly reduced spectrum of sights and sounds for the remainder of their four and a half days' voyage across the Atlantic. Some could find, in the unbroken horizon and the endless play between ship and sea, a relaxing environment that provides escape from the stresses of modern life. For others, the ship's dimensions are all too confining and its vulnerability is all too apparent in the constant and complex motions of the sea. To the technologically inclined passenger, however, the lone ship plying the North Atlantic at twenty-nine knots offers something else again: a near-perfect laboratory in which to reflect upon matters of structure and scale, of engineering and economics.

Before the age of steam, transportation by sailing ship was necessarily slow and subject to the vagaries of the wind, and so regularly scheduled crossings of the ocean were not practical. In the early nineteenth century, for example, traversing the Atlantic took anywhere from three weeks to two months. The earliest steam ships were actually powered by a combination of steam and sail, being driven by paddle wheels only when the sails were useless.

In 1819 the *Savannah,* loaded only with coal and ballast, left her home port in Georgia. She reached Liverpool almost four weeks later, but she had been under steam for a total of only 85 hours because of the quite limited amount of coal such a small ship could carry. Valuable cargo and passenger space was not readily sacrificed for fuel and machinery, and the conventional wisdom was that no vessel would ever be able to store enough fossil fuel to run its engines for the entire journey. In 1833 the 800-ton *Royal William* did carry 324 tons of coal, 8 passengers, and some cargo from Nova Scotia to London in 25 days. However, because she used seawater to make steam, the *Royal William* had had to proceed under sail every fourth day while salt deposits were scraped from her boilers.

During a protracted stay in London in the early nineteenth century, an American lawyer named Junius Smith got involved in importing and exporting. The uncertainty of ship arrivals, and hence their unloading, loading, and departure, made it especially difficult to deal in fresh produce, and at first Smith saw little business advantage in displacing barrels of apples with steam boilers. However, after he personally experienced a fifty-four-day crossing to New York against prevailing westerlies in 1832 and in the following year an even longer return voyage to London, he resolved to establish a steamship company. According to an 1835 prospectus, Smith's plan was straightforward:

> to form a Line, composed of Two British and Two American Steam Ships of 1,000 tons each, which will be sufficient to keep up a communication twice a month to and from New York.... Four steam ships will make as many passages in twelve months as eight sailing ships; and the investment will not equal the cost of eight sailing ships of equal tonnage.

Smith was unable to sell a single share, however, because his proposal flew in the face of the conventional design wisdom, which defined the prevailing technological paradigm. As articulated by Dionysius Lardner, a contemporary British authority on steam engines, the "utmost limit" economically for a fully steam-powered voyage was 2,550 miles, making a transatlantic voyage "perfectly chimerical." He ridiculed those who talked of a direct crossing under steam, saying "they might as well talk of making a voyage from New York or Liverpool to the moon."

However, by this time the conventional wisdom had also come under some more fundamental engineering scrutiny. Although it may have been true that smaller ships did not have enough hull space to carry all the fuel they might need for a fully steam-powered ocean crossing, the fact that such space increased as the third power of a ship's size, whereas resistance rose only as the square, argued for the practicability and desirability of larger vessels, which would require proportionately less powerful engines. Among the early proponents of this conceptual breakthrough were MacGregor Laird, a Liverpool shipbuilder, and Isambard Kingdom Brunel. Laird challenged Lardner's 2,550-mile upper limit in a pseudonymous letter (signed "Chimera"), providing an argument for a 1,260-ton steamer, and Brunel challenged Lardner after his address. But the conventional wisdom was not to be overthrown by theoretical speculations; only real counterexamples could do that.

Smith's British and American Company contracted with Laird's firm for what was to be a 2,000-ton steamship, the *Royal Victoria*, but Brunel's 1,320-ton *Great Western* was completed first. She was outfitted with a surface condenser, which enabled the steam boilers to use recycled fresh water, thus eliminating the problem of salt deposits. Since Smith did not wish to see the Great Western Steam-Ship Company claim the first all-steam crossing, he chartered and hastily modified for ocean travel the 700-ton *Sirius*, which was termed a coastal packet ship because it carried packets of mail up and down the coast. The lighter *Sirius* easily outsprinted the *Great Western* when she was encountered out on sea trials on the Thames, and a fire that subsequently broke out in the *Great Western*'s boiler room discouraged passengers from going on her maiden voyage.

The *Sirius* left Cork on April 4, 1838, with forty passengers and 450 tons of coal. As she approached New York, the ship encountered headwinds and ran short of coal, but by burning spars, cabin paneling, and resin, she was able to complete the first all-steam crossing of the Atlantic in nineteen days. There was much celebration of the historic event, but the *Sirius*'s record was short-lived. The *Great Western,* which had left Bristol four days later, arrived in New York only hours after the smaller ship. Capable of gaining on her at the rate of two knots, the *Great Western* was able to make the transatlantic crossing in just over fifteen days, albeit with only seven passengers on board.

The age of the great steamships had begun in earnest, and so had the competition between Smith's and Brunel's companies. The focus of the competition soon turned to comfort in addition to speed, and by 1840 the British and American Company had launched the *President*—a plusher, if slightly slower, ship than the *Royal Victoria* (which had been renamed the *British Queen* upon the coronation of Princess Victoria in 1837)—but the new vessel was lost in a storm before seeing a year of service. The tragedy kept passengers away from the line's other ship, and Smith's company never did recover.

The *Great Western* continued to be profitable, however, and Brunel's revolutionary *Great Britain* was launched in 1843. She was the first transatlantic steamship with an iron hull driven by a screw propeller instead of side paddle wheels. Her maiden voyage from Liverpool to New York was completed in under fifteen days, and the return voyage in just over fifteen days. The ship was plagued, however, by mechanical and navigational problems that were attributable in part to her novel propulsion system and to the effects of the iron hull on compass readings. In 1846 she ran aground in Durndrum Bay, in northern Ireland, fifty miles off course, and the Great Western Steam-Ship Company was forced to sell its two ships.

In the meantime, another steamship line had begun, the North Atlantic Royal Mail Steam Packet Company, which soon came to be known by the more familiar name of Cunard Steamship Company. Its founder, Nova Scotia businessman Samuel Cunard, saw the ocean as a much more hospitable business environment than the land, for starting a steamship line was much less trouble than starting a railroad: "We have no tunnels to drive, no cuttings to make, no roadbeds to prepare. We need only build our ships and start them to work." In keeping with his no-nonsense attitude, Cunard's first liner, the *Britannia,* was not nearly so luxurious as the *Great Western* and the *President,* but Cunard was not counting on plush cabins and grand saloons to attract passengers and make the business profitable. Rather, he secured a government contract worth £60,000 annually to carry the mail between Liverpool and Boston. The *Britannia* was soon joined by her sister ships *Acadia, Caledonia,* and *Columbia* (beginning a tradition of names ending in *-ia*), and the Cunard line established a unique reputation for reliability and safety that continued long after a profit could not be made by carrying the mail alone. To this day passengers aboard the *QE2* are reminded that Cunard has never lost a passenger at sea through the company's fault.

The technological evolution of steamships in the first half of the nineteenth century was, no less than it would be in the twentieth, clearly driven by economics and a host of related extra-technological considerations, such as passenger appeal. Such forces were especially important in determining the size of the ships, which meant that unprecedented problems in structural engineering and naval architecture had to be addressed. As is so often the case, social and economic forces were driving the technology and testing the engineering to effect it.

Cunard's first ships were termed "immense" by the Liverpool press, and yet Brunel's *Great Britain* was "large enough to swallow two of Cunard's steamships," according to historian L. T. C. Rolt. The confidence of the Victorian engineer was so great that within a decade Brunel was to oversee the building of his 692-foot *Great Eastern,* whose sideways launch was dictated at least in part because of fears that she might break in two if supported at only her bow and stern. (This concern was not a new one. In 1638, as he developed his new science of strength of materials, Galileo had discussed in his *Dialogues Concerning Two New Sciences* the fact that the weight of an object increases as the cube of its size, whereas the resistance to fracture increases as the square. Larger ships could be built safely, his results indicated, only if strictly geometric proportioning was abandoned: To double the size of a ship, one had to more than quadruple the load-bearing section of its hull.)

The difficulties encountered in launching the *Great Eastern*, coupled with its failure to be profitable carrying passengers, left it as the largest ship to be built for decades. Only toward the end of the nineteenth century, amid the renewed optimism on display at the World's Columbian Exposition, were gigantic steamships again being considered seriously. Historian Henry Adams was among those who viewed the exhibits in Chicago in 1893 and, as related in his autobiography, he reflected upon one display in particular:

> Historical exhibits were common, but they never went far enough; none were thoroughly worked out. One of the best was that of the Cunard steamers, but still a student hungry for results found himself obliged to waste a pencil and several sheets of paper trying to calculate exactly when, according to the given increase of power, tonnage, and speed, the growth of the ocean steamer would reach its limits. His figures brought him, he thought, to the year 1927.

Adams's prediction was to prove amazingly accurate. The *Great Eastern* was still the ship of record at the time of the World's Columbian Exposition, to be exceeded, but not by much, in 1907 by the 30,000-gross-ton and 790-foot-long *Mauretania*, sister ship to the ill-fated *Lusitania*. Powered by steam turbine, the *Mauretania* carried 1,750 passengers and could reach twenty-five knots, which enabled it to capture and then hold the transatlantic speed record for more than twenty years.

Early in the new century, the growth of immigration to the United States provided a very lucrative steerage class of passengers to fill the lower decks of ships, and so steamship companies began to realize the liners that had been on the drawing boards. The *Leviathan*, built in Germany as the *Vaterland* but seized and renamed in New York in the midst of war in 1917, was a 50,000-ton, 950-foot-long vessel designed to carry 3,000 passengers.

For a decade or so after World War I, the size of the largest steamships was pretty much fixed at that of the *Leviathan*, and the sinking of the *Titanic* (at 66,000 tons and 880 feet long) in 1912 certainly gave little incentive to build larger. Nevertheless, by the end of the 1920s steamships were carrying more than a million transatlantic passengers per year, and renewed technological confidence and commercial optimism combined to push a new generation of large vessels off the drawing boards. The *Normandie* and the *Queen Mary* were launched in 1935 and 1936, respectively, and each of the 82,000-ton and 1,000-foot-long ships could carry about 2,000 passengers in style. Unfortunately, this new technological scaling up coincided with the Depression. With hard times, the number of transatlantic passengers declined by 50 percent, leaving many a steamship company looking for ways to stay afloat financially. Cruises for the wealthy augmented the traditional ferry service between Europe and America, and carrying the mails (and some light cargo) once again became an important source of revenue. At twenty-nine knots, the newest and largest steamships were much preferred over slow freighters, and medium-sized liners (at 30,000 tons and 700 feet, on the order of the *Great Eastern*) built during this period were frequently called "mailships," thus recalling the economic origins of ocean liners in the packet ships of a century earlier.

World War II placed unique strains on transatlantic shipping. The *Queen Mary* and the *Queen Elizabeth,* the latter launched in 1940, were called into service as troop transports, carrying as many as 15,000 Allied soldiers at a time from New York to the European Theater. This was seven or eight times the normal passenger capacity, and the liners had to be gutted of their luxury, making their innards reminiscent of the days of peak immigration and steerage-class crowding. The ships were refitted after the war, of course, and a dozen or so new liners were launched to handle the new wave of immigrants, refugees, and repatriates that contributed to making the decade beginning in 1948 the most profitable in ocean-liner history.

The *United States,* launched in 1952, was just short of 1,000 feet but displaced only 53,000 tons, in part because of the extensive use of aluminum alloys. She was capable of sustaining thirty-three knots, which enabled her to cross the ocean in just three days and ten hours, thus capturing the coveted Blue Riband, traditionally awarded to the ship achieving the shortest time for a transatlantic crossing. Built to carry just under 2,000 passengers, the *United States* was also designed to be convertible to carry 15,000 troops if need be.

Although World War II reinvigorated ocean liners economically, it had a much more significant technological impact on air travel and thereby indirectly on ocean travel. Out of the experience with jet fighters, the aircraft industry began to develop jetliners to carry commercial passengers. In the early 1950s Britain's de Havilland Comet looked to have the jump on the market, but some early setbacks caused by metal fatigue slowed the introduction of regularly scheduled transatlantic flights. By the time the Comet 4 inaugurated such service in 1958, the reputation of the plane's earlier generations put it at a disadvantage. American-built jetliners proved to be reliable, however, and eventually they captured the confidence of airlines and passengers alike. The growth of air travel signaled the end of the era of great steamships.

The only ocean liner making regularly scheduled transatlantic crossings in the closing years of the twentieth century is the *Queen Elizabeth 2*. Since her final design was on the drawing board when the future of the steamship was already uncertain, the *QE2* was built to serve as a cruise ship as well as an ocean liner. Thus her length and width were chosen to be almost 10 percent less than those of the first *Queen Elizabeth*, so that the new ship, launched in 1967, could pass through the Panama Canal (with barely eighteen inches to spare). As they have in all large engineering projects, matters of inches and pennies have continued to affect technological decisions about the *QE2*. By the early 1980s, the inordinate degree of maintenance and repair required by her steam boilers and turbines, coupled with the increasing cost of petroleum products, prompted a decision to refit the ship with a diesel power plant and electric propulsion system. The waste heat would be used to run flash evaporators to convert enough seawater to fresh for the ship's needs.

The massive conversion project enabled the *QE2* to reach a

peak speed of 33.8 knots during sea trials, prompting speculation that she might recapture for Cunard the Blue Riband held by the *United States*. However, as with many unique engineering projects, there was some unexpected behavior at full scale that was neither suspected nor detected in analytical calculations, scale-model tests, or during full-scale trials.

The larger output of the new power plant demanded that the twin propellers be replaced, and to control the speed of the ship the new design depended upon varying the pitch of the blades while the shaft maintained a constant rate of rotation. Furthermore, to capture some of what was considered waste thrust, freely rotating "pseudopropellers" were installed behind the true propellers. These Grim wheels, named after their inventor, had long been theorized to be capable of providing additional thrust from the tips of their vanes, which were spun by the turbinelike inner part of the vanes, thus increasing the ship's efficiency by as much as 4 percent, a significant advantage on the scale of a superliner. At twenty-three feet in diameter, the Grim wheels were about 10 percent larger than the propellers proper and appeared to work.

The newly engined *QE2* was to be greeted on her second maiden voyage by the princess of Wales, and so skindivers met the ship and inspected her hull as a security precaution before arrival at Southampton. While nothing life-threatening was found, the Grim wheels were discovered to have lost five of their seven blades. New propellers were subsequently fitted, without Grim wheels. The ship now had a maximum speed of thirty-two knots, and hopes of her taking the Blue Riband were no longer considered realistic.

Over the years it has not been the internalist technical details of structure, power, tonnage, or speed that have limited the size and number of ocean liners. Just as steam displaced sail, so the jet airplane has displaced the ocean liner. And the technology of the great Atlantic ships, as embodied uniquely now in the *QE2*, will survive only as long as she is economical to operate.

Among the means the chief engineer employs to reduce operating costs is to compare fuel prices in New York, say, with the price anticipated five days hence in Southampton. The cost of carrying extra fuel is weighed against any potential savings in purchase price, and at the rate an ocean liner uses fuel the pennies can quickly add up to real dollars. Consuming about 370 tons of fuel per day at her service speed of 28.5 knots, the *QE2* can carry enough fuel for twelve days of continuous sailing. At twenty knots the ship can sail for thirty days, which would enable her to complete two-thirds of a circumnavigation of the world without refueling—a feat that would have impressed even Brunel.

But economizing can go only so far on a ship whose selling point is luxury and leisure. In our age of airmail, there is no longer any substitute for passengers to generate revenue. Thus, to attract people to take almost five days to cross an ocean that they can fly across in hours, travel packages of flying one way and taking the ship the other have been promoted by Cunard.

Among the most dramatic pairings of air and sea travel is a supersonic Concorde flight from New York to London with a leisurely *QE2* return crossing. And as disparate as they may seem, these extreme technologies actually share striking similarities in the way their physical size and mode of operation have been shaped by questions of economics and social acceptance. Concorde's takeoff, constrained by the requirements of operating out of JFK International Airport, comprises a carefully orchestrated sequence of turns and bursts of engine power to minimize the sound level over population clusters. The sound barrier can be broken only over open water, a restriction that increases the time required to reach an efficient Mach 2 cruising altitude of 55,000 feet. This stratospheric flight plan in turn requires that the highly pressurized fuselage be small in diameter, thus limiting the number of passengers, which naturally makes for high fares.

The economics of the Concorde are very much concerned with fuel management and consumption, which is especially high at takeoff and landing. To have the required fifteen tons of spare fuel on reaching New York from London, for example, the plane must take off with an extra thirty tons, which is about three times the airliner's payload. To maintain aerodynamic balance of the

Concorde throughout its wide operating range, it is necessary to pump fuel between tanks as the plane accelerates to take off and decelerates to land, so that the center of gravity may be properly positioned. Such sophisticated measures may appear to be unique to a supersonic age, but they are in fact driven by the same economic considerations that nineteenth-century engineers and naval architects wrestled with to design steamships that could carry enough coal to cross the ocean.

Although the aggravated vortex motion that gives additional lift to the Concorde during low-speed flight also causes considerable vibration during takeoff and landing, at supersonic speeds steady flight in the rarefied stratosphere is almost without sensation of movement. Ocean travel, on the other hand, is a much more sensual experience, with the ship being set into constant vibration by its own machinery and by interaction with wind and water. The technologically astute passenger can pick out frequencies of the structure forced by the propellers and the waves, can sense the rigid body motion of the vessel about its center of gravity, and can hear the parts move relative to each other.

On a calm day on the North Atlantic, the sounds of the ship can appear to be the only sounds in the world, and so passengers are understandably curious when a dull thud intrudes. Rumors that a whale or iceberg was hit are soon quashed when the captain announces that the lone ship and the footprint of a Concorde shock wave had just intersected in midocean—an occurrence that seems to the lay passenger, looking out at the seemingly boundless ocean and sky, to be less likely than hitting the jackpot on a slot machine. In the commercial context in which it occurs, such a coincidence of technologies is not really so remarkable. Even though the Concorde is traveling at speeds greater by orders of magnitude than those attained by nineteenth-century steamships, its course, like that of the *QE2*, is still very much determined by the economic considerations embodied in great-circle routes over the North Atlantic.

THE PANAMA CANAL

John Keats's sonnet "On First Looking into Chapman's Homer" is about the poet's sense of excitement and adventure in reading *The Odyssey* in George Chapman's translation. Keats likened to a scientific discovery his experience of finding a vigorous and wholly satisfying English treatment of the much-translated classic Greek adventure story, and he conveyed his excitement in a memorable sestet:

> *Then felt I like some watcher of the skies*
> *When a new planet swims into his ken;*
> *Or like stout Cortez when with eagle eyes*
> *He star'd at the Pacific—and all his men*
> *Look'd at each other with a wild surmise—*
> *Silent, upon a peak in Darien.*

For all its formal and emotional success as a sonnet, Keats's poem is factually flawed in that it confuses Cortez with Balboa, the true discoverer of the Pacific Ocean. But unlike a defective engineering design—in which an incorrectly specified detail can cause the collapse of an entire structure—the poem does not lose its effectiveness because of a historical error, for a sense of discovery is still conveyed and the reader is expected to grant Keats his poetic license. Indeed, to change *Cortez* to *Balboa* would necessitate altering surrounding words, which in turn might require rewriting entire lines, and in the end the whole poem might have to be altered to maintain its scansion, rhyme scheme, and integrity. In this sense a poem *is* like an engineering design.

There are many possible ways to write a poem or to render *The Odyssey* into English, not all of them equally faithful to Homer or history—or equally exciting to modern readers. So also, there have been alternative hypotheses to explain why the planets move the way they appear to do, and alternative designs for virtually every engineering project, be it bridge, building, or canal. But false hypotheses and ill-conceived designs, unlike poems with forgivable flaws, do not endure.

Keats's confusion of Cortez and Balboa, in fact, provides an apt introduction to an engineering story that itself is replete with erroneous hypotheses and wrongheaded design decisions. It is a story that begins in the time of Cortez and Balboa—the tale of the attempts over many centuries first to find a strait and then to construct a canal between the Atlantic and Pacific oceans.

Columbus is credited with discovering South America for the Europeans when he reached Trinidad on his third voyage, in 1498, and then sailed westward along the Venezuelan coast. On his fourth voyage, in 1502-4, he sailed along the Central American coast, where he and many subsequent explorers searched for a strait that they believed provided a natural water passage on to Cathay.

When Balboa discovered the Pacific Ocean in 1513, by crossing the Isthmus of Darién, also known as Panama, the discovery reinforced hopes that a water route between the oceans existed somewhere among the many bays, inlets, and rivers in the region. But the location of the passage continued to frustrate explorers, including Cortez, who looked for a strait as far north as Mexico. With the benefits of modern maps and a bird's-eye view from the window of an airliner, we can see the folly of such a quest. But as one sails along the coast of Central America even today, it is easy to imagine how Columbus and his contemporaries could view the isthmus as but another chain of labyrinthine Caribbean islands whose careful exploration would yield a passage westward to the East.

By 1525, however, after decades of searching, a natural waterway between the oceans was being referred to as "the doubtful strait," and a canal was being suggested as early as 1528. Alvaro de Saavedra de Cerón, who at different times served under both Cortez and Balboa, is generally credited with originating the idea. He no doubt was able to observe firsthand the extreme narrowness of the Panamanian isthmus, and thus identified it as a likely location for a canal project. Increasing settlements in Central America and the growth of trade created plenty of demand for communication between the oceans, but for many years this need was to be served only by the *camino real,* which started out as little more than a mule track. (The Straits of Magellan, discovered in 1519, were too far south and far too dangerous to be of much practical use for regular trade routes.)

Early proposals for a canal included extending the navigability of the flood-prone Chagres River into the Panamanian isthmus, but such an ambitious engineering project naturally depended upon sufficient financial and political support. In 1534, the first survey of a canal route through the Isthmus of Panama was ordered by Charles I of Spain, but for all the arguments in favor of a canal, there was also opposition to any artificial waterway. The objections fell into three categories: (1) Nations other than Spain would benefit from such an easy trade route; (2) breaching the land between the oceans, then thought to be at vastly different levels because the tides were poorly understood, would cause uncontrollable inundations; and (3) divine punishment would fall upon those who might dare, in the words of one critic, "to alter the form which the Creator, with supreme wisdom and forethought, designed for the structure of this universe."

In spite of the real obstacles and threatened consequences, explorations and surveys continued for three centuries throughout the Americas. By 1811 the German scientist and explorer Alexander von Humboldt could identify nine possible locations for canals that would extend existing river routes to cross the land and provide unobstructed water passage between the Atlantic and Pacific. In his assessment, the top four locations were in Central America, in the following order of preference: Nicaragua, Colombia, Mexico (Tehuantepec), and the Panamanian isthmus. Regardless of the choice of site, Humboldt cautioned that no project should begin without a sufficiently thorough survey of all alternative routes and "above all it would be a mistake to undertake such a work on too small a scale; for ... the cost does not increase in the same proportion as the canal section and the breadth of the locks." Among those excited by Humboldt's vision was the aged poet Goethe, who in 1827 thought it "absolutely indispensable for the United States to effect a passage from the Mexican Gulf to the Pacific Ocean," and regretted that he would not live fifty more years to see the accomplishment.

No canal, large or small, was to be completed for almost another century, but there were to be many false starts. With an engineering project the scale of a transisthmian canal, the decision as to site is only one of many that must be made before detailed engineering planning and design can begin. Among the principal nontechnical obstacles to commencing were those relating to economics and politics. In an area not known for political stability, such concerns as sources of capital, concessions to build across the countries controlling the land, and guarantees of long-term monopolies and security for a completed canal, were as important as those relating to the type of canal to be built.

At the middle of the nineteenth century there still was no canal, but there was a growing transisthmian commerce. A railroad was completed from coast to coast in 1855 and thus became the first transcontinental railroad in the Americas. Contrary to most people's sense of geography, the railroad (and the canal that would eventually follow roughly the same route) actually proceeded generally in a southeasterly direction in traversing the isthmus from the Atlantic to the Pacific. In spite of this counterintuitive way to get from east to west, the Panama Railroad proved to be a boon to travelers who wanted to get from New York to California. Ship passengers would disembark at Limón Bay for the forty-eight-mile railroad ride to the Pacific Coast, where they would board a second ship, at Panama Bay. This was such a time-saving route that railway tickets commanded premium rates, and it took $5 in gold just to purchase the right to walk across the isthmus along the railroad tracks.

Railroads developed so rapidly during the middle decades of the nineteenth century that in 1880 James Buchanan Eads, who had already bridged the Mississippi at St. Louis and constructed jetties at the river's mouth to maintain an adequate channel depth, proposed a "ship railway" across the Isthmus of Tehuantepec. According to Eads's scheme, fully laden oceangoing vessels (then of the order of 7,000 tons) would be lifted out of the water at one terminus, cradled on many-wheeled platforms, and pulled by powerful locomotives on several parallel tracks to the other ocean. This would save the time, effort, and expense of transferring cargo to and from conventional transisthmian railroad cars. Eads's idea was not so fanciful as it might seem, for canal boats had long been hoisted up inclines, and some had even been transported by rail across the Allegheny Mountains. But the scale of the Mexican plan, although it was projected to cost only a fraction of other transisthmian ship canal schemes, required some financial backing from the government. Other proposals current at the time included canals across the Panamanian and Nicaraguan isthmuses, and debates over technical issues and lobbying over political and economic ones put off time and again any final decision.

Science and scientists, engineering and engineers, were at heart not much different in the nineteenth century than they are today. In response to the obvious desire for and practicality of a canal across Central America, societies began to form to study the problems associated with such a project, and meetings were organized. In 1876 a committee of the Société de Géographie de Paris invited geographical societies from around the world to be represented at an international convention and to share the expenses of a careful exploration, possibly employing balloons for aerial observation, so that a definitive canal route could finally be chosen. In the meantime, a private company, the Société Civile Internationale du Canal Interocéanique du Darién, was formed, and it volunteered to shoulder the expense of the exploration. The first report to the company described options, including a tunnel through the continental divide and a canal with locks.

An international congress was held in Paris in May 1879 to reach a conclusion regarding the various options. The congress was chaired by Ferdinand de Lesseps, a septuagenarian Frenchman who was neither engineer nor financier but has been described as "a promoter whose ambition was not the accumulation of wealth but the accomplishment of tasks called impossible by lesser men." De Lesseps was "one of the most eminent personages in the world" at the time and was known simply as *le grand Français*. His achievement in building the Suez Canal had given him a supreme confidence that he could also direct the construction of a canal across Central America, and he orchestrated the Paris congress accordingly.

At the first plenary session, held in the auditorium of the Société de Géographie, the delegates heard a clarion call to action: "The cutting of a canal through Central America has now become essential to the welfare of all nations.... The building in which we are met is dedicated to science, and the impartial serenity of science will impress itself upon your deliberations." De Lesseps implored the congress "to conduct its proceedings in the American fashion, that is with speed and in a practical manner, yet with scrupulous care." The delegates divided into five committees, to consider such questions as volume of canal traffic, width, tolls, and so forth, but many of their recommendations would really depend upon the endorsements of the largest and most important committee, that charged with considering the defining aspects of the macroengineering enterprise: choice of site, type of canal, and cost. In the course of two weeks of delegates being wined and dined around Paris, the debate over alternative routes took place.

A rational decision depended upon a consideration of such matters as whether rainfall was sufficient to support the operation of locks. The least expensive and most easily completed design appeared to be a lock canal across Nicaragua, taking advantage of an existing lake. But the influential de Lesseps was so opposed to the inclusion of locks that the technical committee rejected all such proposals. In spite of the uncertainty of cost due to questions of stability of the slopes that had to be cut, and of control of the notoriously uncontrollable Chagres River, de Lesseps was decidedly for a sea-level canal between the bays of Limón and Panama. His dominant presence at the closing plenary session carried the vote for his personal and emotional, rather than technically sound, choice. Only nineteen of the seventy-four affirmative votes for a sea-level canal were cast by engineers, and most of those supporting de Lesseps had been associated with the Suez Canal project. As for de Lesseps himself, he had won the day by declaring, "I vote 'yes'! And I have accepted the leadership of the enterprise."

Within months of the endorsement, de Lesseps began the uncertain task of raising capital, and in 1880 actual work on the project began. But this canal proved to be quite a different proposition from Suez. As one observer wrote of the Panama project in 1886: "There probably was never a more complicated problem—a problem embarrassed by a larger proportion of uncertain factors—presented to an engineer.... Every step...is more or less experimental." There was much rock, whereas there had been mostly sand and soft soil at Suez; there were elevations as much as 330 feet to cut through, as opposed to the maximum 50 at Suez; there were no nearby population centers from which to draw cheap labor; there were the floods of the Chagres River; there was the humidity to add to the heat of Suez; and there were also the deadly fevers.

The French experience proved to be a disaster, largely because of de Lesseps' stubborn refusal until 1887 even to consider a redesigned canal containing locks, and then only as a temporary solution. In the meantime, repeated landslides had made it clear that substantially greater volumes of earth and rock had to be excavated to achieve stable slopes through the isthmus. By the end of the decade the company was in financial trouble and unable to raise badly needed capital. There were scandals regarding excessive payments and outright bribes to reporters to promulgate stories of success when in fact the project was replete with failure. Among those implicated were Gustave Eiffel, an early opponent of a sea-level canal, whose later contract to construct locks and the attendant machinery was considered "overgenerous." The canal company ended up in receivership.

After five years of inactivity, a second French company was formed, but it too failed to achieve success. Repeated landslides that required more and more excavation, and the illnesses that took their toll among the workers, continued to frustrate French efforts. The landslides had a very visible source and could eventually be overcome by sheer patience and hard work. The fevers, however, were more insidious, and were to claim the lives of as many as two of three workers.

Although it was hypothesized as early as the midnineteenth century that the mosquito transmitted yellow fever, the conventional scientific wisdom was that decaying matter in the soil and generally unsanitary and unhygienic conditions were the true cause. French hospitals in Panama were held up as models of cleanliness and attractiveness. Wards were brightened up with tropical flowers, and to keep the voracious ants away from them the pots were placed in bowls of water. This was, of course, a very efficient means of breeding mosquitoes, and the hospitals were in fact inadvertent death traps. By the end of the century the twenty-year French effort came to an end, and the company's excavations and equipment were up for sale.

The familiar palindrome "a man, a plan, a canal, Panama" places the engineering achievement before the political entity, but in fact it was the junta that created the Republic of Panama which finally determined the fate of the canal. There was growing support in the United States for an American-controlled canal. The Nicaraguan route had strong adherents because of the prior claim of the French to the Darién isthmus, because of later difficulties in getting Colombia to agree to terms for granting the United States a concession across the Isthmus of Panama, and because of Nicaragua's supposed environmental advantages. One of the engineers for the French project, Philippe Bunau-Varilla, in the end proved to be the agent who shifted sentiment in Washington to the route more distant from the States. As a crucial vote approached in the Senate, Bunau-Varilla recalled that Nicaragua had once issued some postage stamps depicting the eruption of Momotombo, and he bought up enough of the stamps from Washington collectors and dealers to send a copy to each senator just days before a crucial vote authorizing the purchase of what the French had left behind. Needless to say, assurances that Nicaraguan volcanoes were dormant, and thus posed no threat to a canal there, were of little avail.

With the French equipment, maps, and experience, the Americans would be able to attack the isthmus with renewed vigor and resources—if an agreement could be reached with Colombia as to compensation for canal rights through the country. This was not forthcoming, and the United States did not at all discourage the revolution that created the Republic of Panama. Still, a Nicaraguan site continued to be held up as a possibility until the French agreed to accept much less than they wished for their Panamanian legacy. The deal was finally closed in 1903, for a price of $40 million.

The Panama Canal has been called the "greatest liberty Man has ever taken with Nature" and an "unparalleled engineering triumph." Whichever view is taken, the achievement would not have been possible without many economic and political machinations and the determination and dedication of key individuals, many of whom were necessarily engineers on the great project. The American effort began in earnest in 1904, with John Findley Wallace, who had had extensive experience building railroads in the Midwest, as chief engineer. The chief sanitary officer was William Crawford Gorgas, who had experienced much success in eradicating yellow fever and malaria in Cuba. Neither had free rein, however, as a seven-man commission directed operations from Washington, under the supervision of Secretary of War William Howard Taft. Wallace eventually convinced Taft that a reconstituted commission was needed, but shortly thereafter the two men had a falling-out and Taft demanded Wallace's resignation.

Within three days John Frank Stevens, a distinguished railroad engineer, was persuaded to be Wallace's successor, after President Theodore Roosevelt agreed to Stevens's conditions that he be un-hampered by red tape and have virtually absolute authority in making decisions about the canal. Furthermore, he promised to stay in the post of chief engineer for the canal only until he "had made its success certain, or had proved it to be a failure." On his arrival in Panama, Stevens found a lack of organization and ini-tiative, and he began immediately to turn things around. Al-though the canal had been under construction off and on for twenty-five years, it was what today would be called a "fast track" project, in that not all final design decisions had yet been made. In particular, the debate over a sea-level versus a lock canal had not been settled. In 1905 Roosevelt appointed a board of con-sulting engineers to make that decision once and for all. With Stevens's strong input, a lock canal was finally decided upon.

The scheme adopted was very similar to one proposed, but ig-nored, at the 1879 Paris Congress by Godin de Lépinay, one of the few French engineers who had any extensive work experience in the American tropics. His idea had been to dam the Chagres River, thereby creating a gigantic man-made lake about eighty feet above sea level. Ships would be lifted via locks from the At-lantic Ocean to the lake, across which they could sail toward the continental divide. The major excavation would take place there, and a steep- and high-sided channel would allow the passage of ships to another series of locks, which would lower them to the Pacific. Had the congress adopted such a design, de Lesseps and the French might have succeeded in building a canal. Not only did de Lépinay's concept minimize the amount of excavation, but it also provided a means of controlling the floodwaters of the Chagres while at the same time providing a source of freshwater to fill and empty the locks, using gravity instead of pumps.

With a firm decision finally made as to a canal design and with Stevens in place, progress soon became apparent. Much of it was methodically recorded by Ernest Hallen, the official photogra-pher of the canal commission, and visiting journalists. Indeed, the

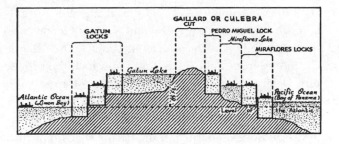

*Vertically exaggerated cross section of the Panama Canal,
showing change of elevation achievable with locks*

canal was one of the first major news stories whose reporting
benefited from advances in photomechanical printing processes
made in the late nineteenth century, which obviated the laborious
conversion of photographs to line drawings before they could be
printed. As a result, the plethora of newspaper stories, magazine
articles, and books that appeared during the American construction phase were richly illustrated.

One of the most compelling images and arduous aspects of the
great construction task was the excavation of the Culebra Cut
through the continental divide. An efficient system of removing
earth and rock and disposing of it was particularly critical, and
Stevens made wise and decisive choices even amid design uncertainty. Late in 1906, Roosevelt visited Panama, becoming the first
U.S. president to leave the country during his term in office. He
was indefatigable in his tour of the canal site, which included
everything from posing in the cab of a steam shovel to attending
formal banquets. Shortly after returning to Washington, Roosevelt appointed a new commission. He made Stevens chairman
as well as chief engineer. But Stevens soon reminded Roosevelt
of their agreement and resigned his post, satisfied that the canal
was well under way.

Tired of having civilian engineers quit the project, Roosevelt
sought to "turn it over to the army" and replaced Stevens with

Lieutenant Colonel George Washington Goethals, who took over on April 1, 1907. Although Goethals never appeared in military uniform in Panama, he was a forceful and powerful leader who divided up responsibility for the enormous project. The Atlantic Division, extending from Limón Bay up to and including the locks and Gatún Dam, was put under the direction of army engineer Major William Luther Sibert; the Central Division, which included Gatún Lake and the infamous Culebra Cut, came under Major David Du Bose Gaillard; and the Pacific Division, which extended from the end of the cut to Panama Bay, was overseen by the civilian engineer Sydney B. Williamson.

All divisions had tasks of comparable technical difficulty, but the excavation of the cut through the continental divide overwhelmed all others in sheer magnitude. The French had excavated 78 million cubic yards of earth during a twenty-year period, and from 1904 to 1907 the Americans removed another 14 million. In the final seven years of the project, 219 million cubic yards, or more than two-thirds of the total, were removed under the direction of Gaillard. While the highest cliffs along the canal are composed of firm volcanic rock and are thereby relatively stable, numerous landslides occurred elsewhere along the cut and often undid months of work. However, Goethals, Gaillard, and the "Army of Panama," as Goethals referred to the workforce, attacked each new obstacle methodically. Unfortunately, about a year before the completion of the canal, Gaillard was forced to undergo an operation for a brain tumor and died late in 1913. In 1915, President Woodrow Wilson issued an executive order changing the name of Culebra Cut to Gaillard Cut, in recognition of the engineer's central role in its creation. Gaillard thus became one of the few engineers memorialized by name in Panama, but when engineering school administrators were asked in a 1930 survey to identify the greatest engineers of all time, three of the dozen most frequently named were de Lesseps, Stevens, and Goethals. Which name should be most closely associated with the overall Panama Canal project remains a subject of debate today.

The canal opened officially on August 14, 1914, but the course of events then unfolding in Europe overshadowed the long-awaited opening and mocked the slogan on the Canal Zone's official seal: "The land divided, the world united." The canal was plagued with landslides in the early years of its operation, but after each slide the channel was methodically dredged open again. During World War II the canal was a strategic link between the Atlantic and Pacific fleets, and veterans still recall having cameras confiscated when they would shoot pictures of ships floating high above sea level on Gatún Lake.

The Panama Canal has been described as the technological equivalent in its time of putting a man on the moon, but today some see it as lacking in capacity. The locks normally limit the size of ships that can pass to about 965 feet in length, 106 feet in beam, and 39.5 feet in draft. Ships have even been designed to those specifications so as to allow canal passage. Although highways now provide transcontinental alternatives to the canal, there is still a keen desire to maintain a modern water route between the oceans. A tripartite committee, comprising representatives of the United States, Japan, and Panama, is presently considering alternatives for improving the canal. Among the alternative being looked at is de Lesseps' dream of a sea-level canal, now all the more important because deforestation is affecting the water run-off into Gatún Lake, and there is some concern that the freshwater supply that has allowed the locks to operate solely through gravity flow may be in jeopardy.

Although the Panama Canal, even if superseded, will remain one of the greatest engineering achievements of all time, its more recent history has been dominated by political rather than technical considerations. Under treaties that went into force in 1979 the Panamal Canal Company was replaced by the Panama Canal Commission, an agency of the U.S. government, headed by an American administrator until 1990. Since then, a Panamanian has served as administrator, and on December 31, 1999, the canal and its operation will be turned over completely to Panama. Although a political era will end, ships will no doubt continue for a long time to traverse the canal between Colón (Spanish for Colum-

bus) and Balboa. Regardless of who controls these ports, so fittingly named after the explorers most closely associated with the
Atlantic and Pacific oceans, they evoke the long and rich history
of a transisthmian canal project that in turn makes abundantly
clear how interrelated are the technical, economic, political, and
human aspects of science and engineering.

THE FERRIS WHEEL

Engineering achievements are shaped as much by their times as by the technical state-of-the-art of those times. Such is the lesson taught by the intertwined story of two of the best-remembered achievements of a century ago: the Eiffel Tower and the Ferris wheel.

When the 1889 Paris Universal Exposition was in its planning stages, a consensus developed that a very high tower would be a fittingly distinctive monument to commemorate the centennial of the French Revolution. Obelisks had long been used as monuments, of course, but in the nineteenth century it became common for engineers to propose iron towers taller than was practicable for stone monoliths. In 1833 the British engineer Richard Trevithick, who had early in the century developed the first steam locomotive to run on tracks, had proposed a cast-iron tower a thousand feet high in commemoration of the Reform Bill that Parliament passed the year before. Another structure of similar height was put forward as a fitting reuse of the disassembled iron-and-glass components of the Crystal Palace, which housed the first world's fair, in London in 1851. It was in such a tradition that two engineers who worked for Gustave Eiffel's firm, Maurice Koechlin and Émile Nouguier, sketched in May 1884 yet another tall tower.

Eiffel's engineering firm had significant experience in bridge building and had recently completed the structural framework

for the Statue of Liberty. Thus, after carrying out some basic calculations, the Eiffel engineers knew definitely that their four-
legged pylon could be constructed out of riveted wrought iron
for the upcoming Universal Exposition. Eiffel was not overly enthusiastic when he was shown some of Koechlin's first sketches.
He did authorize his engineers to continue studying the problem,
however, and he embraced the idea after an architect with the
firm, Stephen Sauvestre, made further sketches that reduced the
number of cross members but embellished the pylon with monumental arches and other ornamental details. In September, Eiffel took out a patent with his engineers, whose rights he
subsequently bought out. The tower that had become Eiffel's became the centerpiece of the exposition.

Not everyone was thrilled about having Paris defaced with
what could effectively be described as a bridge to the sky cantilevered up from the ground. The artistic and literary community,
in particular, protested "in the name of French art and history"
against "this offense to French good taste." An 1887 article in *Le*

*The Eiffel Tower drawn to scale
among famous monuments*

Temps, signed by Guy de Maupassant and Alexandre Dumas, among others, described the rising tower as the product of the "baroque, mercantile imaginings of a machine builder." Eiffel responded forcefully, defending the tower as "beautiful in its own right." And he continued:

> Can one think that because we are engineers, beauty does not preoccupy us or that we do not try to build beautiful, as well as solid and long lasting structures? Aren't the genuine functions of strength always in keeping with unwritten conditions of harmony? . . . Besides, there is an attraction, a special charm in the colossal to which ordinary theories of art do not apply.

The bare tower of bridge girders, as conceived by Koechlin and Nouguier, might not have been so readily defended. Had it been selected and built, an unlikely event in the face of the criticism, it might have been destroyed shortly after the Paris Universal Exposition had closed. The success and longevity of his tower must be attributed to Eiffel's genius, for he knew not only what would work structurally but also what would be a social, political, financial, and symbolic success. Almost two million people visited the tower during the fair, bringing in receipts that all but equaled the construction cost of 7.5 million francs, and the tower turned a profit shortly after the fair closed.

Indeed, the Eiffel Tower was such a success that it had already become the symbol to be overshadowed when the World's Columbian Exposition was being planned for Chicago to celebrate the four-hundredth anniversary of Columbus's landing in America, albeit a year late, for the exposition was to be delayed until 1893. The architect Daniel H. Burnham was appointed chief of construction for the project, and his partner, John W. Root, consulting architect. Together, Burnham and Root had gained fame as the designers of the ten-story Montauk Building, Chicago's first tall edifice and the one to which it is said that the term "skyscraper" was initially applied. Their Masonic Temple, to open at the corner of State and Randolph streets in 1894, would then be the tallest building in the world. When Root died suddenly, early in 1891, Burnham was left alone with the responsibility of organizing the design and construction of the fair complex that would be called the White City, after the uniform color of the buildings and their illumination at night by the greatest use of electricity in the nineteenth century.

Early on in the planning of the exposition, the architectural decision was made that all the major buildings surrounding the Court of Honor would have a uniform cornice line and all buildings at the fair would follow a classical theme of "Greco-Roman-Oriental style." Up until that time, no classically styled buildings had been erected in Chicago, and so the decision was not a universally popular one among architects and critics, but once made, it was pursued with vigor in the interests of expediency. When he would look back at the exposition in his third-person autobiography, Henry Adams would give the architecture of the "trading city" a backhanded compliment: "All trader's taste smelt of bric-a-brac; Chicago tried at least to give her taste a look of unity."

While construction progressed on the exhibition buildings, no progress had been made in finding some truly novel structure for the Chicago fair that would "discount" the Eiffel Tower. As one contemporary reporter described the situation subsequently, "American pride was at stake." The young country promised "the greatest exposition that had ever been held on earth." America had already established its industrial achievement, and the fair was to be its opportunity to demonstrate that its ingenuity was also equal to that possessed in the Old World.

It was in such an atmosphere that the suggestion that the fair be all under one roof had earlier brought a proposal for a 3,000-foot-diameter tentlike structure with a 1,000-foot-high center pole. Grand schemes had abounded; one had involved buying the Roman Colosseum and reerecting it in Chicago. As can be imagined, there were many proposals for towers greater than the Eiffel, one being 1,500 feet tall with an almost 500-foot-diameter base. Inside this monumental structure would have been 5,000 hotel rooms and a 237-foot-diameter dome that would cover a music hall. But even having the greatest of towers ever would have left the fair open to the criticism of being derivative. According to one observer, writing after the fact, the only option left to American ingenuity was to "set the Eiffel tower on a pivot and put it in motion."

No such thing had been put forth as late as 1891. When Burnham found himself at a banquet addressing architects and engineers, he praised the former but excoriated the latter for not having met the expectations of the people. Nothing had been proposed that displayed originality or novelty to rival the Eiffel Tower. He wanted something new in engineering science, but Burnham felt the engineers were giving him only towers.

Among the engineers at the banquet was the youngish George Washington Gale Ferris, Jr. He was born in Galesburg, Illinois, in 1859, and at age five moved with his family to western Nevada. There, while living on a ranch, he became fascinated with a large undershot water wheel, which raised buckets of water out of the Carson River to supply a trough for horses. Ferris would later recall his fascination with the wheel's action, but, according to some accounts, as a youngster he was not equally fascinated with formal education. He was sent to military school in Oakland, California, and then attended Rensselaer Polytechnic Institute, from which he graduated with a civil engineering degree in 1881. After some early work involving railway, mining, and tunnel engineering, he became involved with bridge building, and by 1885 had responsibility for testing and inspecting iron and steel made in Pittsburgh for the Kentucky & Indiana Bridge Company. At the time, steel was just being introduced into bridge building, and Ferris saw an opportunity for a new business. He organized in Pittsburgh the firm of G. W. G. Ferris & Company, which conducted mill and shopwork inspection and testing for others. Soon the firm was involved in one way or another with much, if not most, of the steel bridge construction going on in the United States. Thus, by 1891, Ferris had considerable practical experience in the specification, fabrication, testing, and erection of steel structures.

When Ferris would later be asked where the idea for his great wheel came from, he recalled that, a while after hearing Burnham's challenge, he found himself at a Saturday afternoon dinner club made up mainly of world's fair engineers. According to Ferris,

> I had been turning over every proposition I could think of. On four or five of these I had spent considerable time. What were they? Well, perhaps I'd better not say. Any way none of them were satisfactory.... It was at one of these dinners, down at a Chicago chop house, that I hit on the idea. I remember remarking that I would build a wheel, a monster. I got some paper and began sketching it out. I fixed the size, determined the construction, the number of cars we would run, the number of people it would hold, what we would charge, the plan of stopping six times in the first revolution and loading, and then making a complete turn,—in short, before the dinner was over I had sketched out almost the entire detail, and my plan has never varied an item from that day. The wheel stands in the Plaisance at this moment as it stood before me then.

Perhaps not surprisingly, because it was so novel, Ferris's proposal received mixed reactions. The directors of the world's fair first granted Ferris a concession to build his wheel, then withdrew their approval, and then gave the go-ahead again, on December 16, 1892. The fair was to open on May 1, 1893. He had to raise $350,000 and locate, fabricate, ship, and assemble 2,100 tons of material in a matter of months.

Ferris's familiarity with the Pittsburgh iron and steel industry—along with his experience in testing, inspection, and construction—enabled him to expedite the fabrication of his wheel's components, have them made correctly the first time, and have everything shipped as quickly as possible. Various steel mills provided the material, which was assembled into components in Detroit and shipped to Chicago in 150 railroad cars in late March. Among the components was the axle—the largest steel shaft ever forged. This centerpiece of Ferris's wheel alone weighed about 45 tons and was more than 45 feet long and 32 inches in diameter.

When the parts began arriving in Chicago, the foundation for the superstructure was ready. It had been started in January, when eight 20-foot-square holes 35 feet deep were dug to receive the concrete to which the steel legs would be bolted. After the massive axle was hoisted into place, the wheel could be assembled. The order and method of assembly of such a large and novel structure was crucial, of course, and the responsibility for its success fell on Ferris's partner, William F. Gronau, a fellow Rensselaer engineer, who most likely did the structural design calculations. The wheel was essentially a gigantic bicycle wheel 250 feet in diameter. Whereas the passenger cars would be carried on a relatively stiff trusswork rim, the rim itself would be connected to the axle by 2½-inch-diameter rods that acted like spokes, which would be incapable of carrying any significant compressive load. Thus the structure was properly termed a tension wheel, and the largest wheel of that design theretofore constructed was a 35-foot example in Scotland. Ferris would no doubt have been familiar with a 60-foot Burden water wheel that stood in Troy, New York, not far from his alma mater. The largest wheel of any kind in 1892 was a 72-foot-diameter overshot water wheel on the Isle of Man.

For all the novelty of its scale, Ferris's wheel did have these and other antecedents, as do all artifacts. "Pleasure wheels" existed in Bulgaria in the early seventeenth century, as the English traveler Peter Mundy recorded in his diary in 1620. He accompanied a sketch with this description: "Children sitt on litle seats hunge round about in severall parts thereof, And though it turne right upp and downe, and that the Children are sometymes on the upper part of the wheele, and sometymes on the lower, yett they alwaies sitt upright." Not surprisingly, pleasure wheels moved westward, and by the 1870s in America there were steam-driven wooden and metal pipe models of the order of 50 feet in diameter, some carrying upwards of 50 passengers.

As do all structures that are novel in principle or size, the Fer-

An eighteenth-century pleasure wheel

ris wheel attracted its naysayers. Some said it would never turn, others that it would keel over in the wind, and still others that it would become unbalanced by its passengers. When it was officially opened on June 21, 1893, the wheel turned perfectly and continued to do so for its lifetime. When Ferris saw a storm coming in July, he, his wife, and a reporter boarded the wheel to ride out a 110-mile-per-hour wind. According to the reporter, "The inventor had faith in his wheel; Mrs. Ferris in her husband. But the reporter at that moment believed neither in God nor man." To the concern about its becoming unbalanced, Ferris responded, "The ego in mankind makes it hard to realize that the people on this great wheel are hardly more than so many flies." Hyperbole, indeed, but even when fully loaded the wheel itself outweighed its passengers by a factor of ten.

Novel engineering design achievements soon become the objects of study of engineering scientists, and shortly after the Ferris wheel was turning so were the gears of analysts' minds and the presses with technical papers about it. In the March 24, 1894, issue of *Engineering News*, J. W. Schaub wrote of the "conjectures as to the probable stresses which may occur in such a structure" and of the "great disappointment that the subject was not handled more scientifically" by Ferris: "For some time engineers have been waiting for a paper from the designer of the wheel, giving a full description of its design, but it does not appear that such a paper is forthcoming." Indeed, it was not, but publication can be as far from the mind of some designers as design is from that of some analysts. To many an inventor, the artifact speaks for itself. That is not to say that no analysis goes into a design, for it surely does, whether on a chophouse table or on a drafting board, but the idea of writing the details up for publication is as alien to some engineers as explaining their paintings is to some artists. In Ferris's case he may not have wanted to share the glory with his partner, Gronau, or it may simply have been that Ferris had the great preoccupation of having lost his fortune soon after the initial success of his wheel. Worry over business matters was blamed for his coming down with typhoid fever and dying at age thirty-seven on November 22, 1896.

Reading an analysis of the Ferris wheel may also have been of little interest to the nearly 1.5 million fairgoers who rode it, but they did revel in its statistics and logistics. Each of the 36 cars was the size of a trolley car and had 40 swivel stools, but some passengers also stood. At a total of 60 persons per car, the capacity of the wheel was 2,160, and as many as 38,000 people rode the wheel in a single day. It has been estimated that the Ferris wheel made about 10,000 revolutions during the 19 weeks it operated on the Midway Plaisance, the concession and amusement area of the fair that gave its name to midways at all subsequent fairs. Uniformed guards rode in the cars for safety and to facilitate loading and unloading, which was done through separate doors. Six cars were loaded at a time, which required the wheel to make the six stops that Ferris had prescribed for taking on a totally fresh load of passengers. Once the cars were loaded, the wheel would make one complete revolution, and so each passenger would experience at

*The Ferris Wheel on the Midway Plaisance at the World's
Columbian Exposition, 1893*

least two revolutions during the 20-minute ride. At 50 cents per ride, as opposed to a nickel for the merry-go-round, Ferris's wheel was a short-term financial success.

When the World's Columbian Exposition was over, the disposition of the Ferris wheel presented a new problem. It would have cost $150,000—a huge sum at the time—to move and rebuild the wheel in New York City, as was once proposed, and so that never came to be. The great wheel stood immobile over the deserted fairgrounds throughout the winter of 1893–94. The following spring, its parts were disassembled, numbered, and loaded onto railroad cars, with the special car designed to carry a massive Krupp gun reserved to transport the 70-ton axle with attached hubs. The cars did not move, however, and the wheel's parts remained on a railroad siding throughout the following winter. Only in the spring of 1895 was the wheel reassembled on Chicago's North Clark Street, but the small amusement park surrounding the great wheel could not draw enough visitors to support it. Ferris's wheel was reerected for the last time at the Louisiana Purchase Exposition in St. Louis in 1904, after which it was abandoned there. The rusting eyesore was finally dynamited in 1906, and rumors persist that the great axle still lies buried in Forest Park.

Although the original Ferris wheel thus met an inglorious end, it did give rise to many imitators, which also bested it in various ways. In 1894 the Firth wheel was erected at a San Francisco exhibition. Though it was only 100 feet in diameter, the elevation of the ground on which the wheel stood gave its riders a spectacular view from more than 300 feet above the Pacific Ocean. Several wheels to rival and better Ferris's were erected in the late 1890s by Walter B. Basset, a retired British naval officer. His first was the 270-foot-diameter Great Wheel, which went up in London for the Oriental Exhibition of 1895. Among its attractions were cars reserved for smoking and nonsmoking passengers, and a hollow axle through which visitors could walk, for an additional fee. The Gigantic Wheel erected at the seaside resort of Blackpool featured at one time a car equipped with a Ping-Pong table and was advertised as having something for everyone:

> *The most beautiful Iron-work Construction in the North of England.*
> *Engineers should see it!*
> *The most lovely panorama in the world.*
> *Artists should paint it!!*
> *The most interesting snap-shots.*
> *Photographers should snap it!!!*
> *The finest air in the universe.*
> *Doctors should recommend it!!!!*

Basset erected a 300-foot wheel, La Grande Roue, for the Paris Universal Exposition of 1900, but it was dismantled in 1920. Basset's 197-foot wheel, known as the Riesenrad, was built in Vienna in 1897 and rebuilt after being damaged during World War II. It was the world's largest operating Ferris wheel until 1981, when a 208-foot model was built in Japan. Today the largest is the Giant

Giant Ferris wheel proposed for London

Peter, whose 279-foot-diameter wheel turns over Himeji Central Park in Hyogo, Japan. In 1996 it was announced that a 500-foot-diameter wheel would be erected beside the River Thames in London to mark the turn of the millennium.

Ferris wheels have also spun off many related amusement-park rides. Although none of them necessarily stands as high engineering, each descendant does show once again how artifacts from paper clips to steamships evolve by removing some real or perceived failure of their ancestors to achieve unqualified success. Since one of the greatest shortcomings of the largest Ferris wheels was the high cost of putting up, taking down, and moving them, the bridge builder William E. Sullivan concentrated on developing a more portable model. He erected his first 45-foot-diameter wheel in a park in Jacksonville, Illinois, in 1900. He later founded the Eli Bridge Company in Roodhouse, Illinois, and advertised that the "New Big Eli Wheel" was booking orders for 1906. The "BIG WHEEL when up, the LITTLE WHEEL when down" could easily be erected or dismantled and ready to move to another fair site in a matter of hours, as opposed to the weeks it took to ready Ferris's wheel. In 1919 the Eli Bridge Company relocated in Jacksonville, on both north–south and east–west rail lines, and it remains today a major manufacturer of Ferris wheels.

Although they continue to be a staple of midways, Ferris wheels may be considered tame and boring rides by today's standards, and in traditional models the whole ride must be stopped frequently to unload and load a few passengers at a time. To remove such an objection, the Aerio Cycle erected at the Pan American Exposition of 1901 in Buffalo had one small wheel at each end of a seesawlike beam, so that half the passengers could revolve high above the midway while others were being loaded. Joseph Strauss, chief engineer of the Golden Gate Bridge, applied his experience with bascule bridges to design and patent the Aeroscope, which raised 118 people about 200 feet above the 1915 Panama-Pacific International Exposition and turned about its base to give everyone a panoramic view of San Francisco. The ride had a single car as big as a two-story house, so that all passengers were up in the air at the same time.

Sixteen of the twenty-four cars on Coney Island's 135-foot Wonder Wheel are mounted on tracks so that they move in and out each time the wheel makes a revolution. Another ride, the Zipper, though it can load only two cars at a time, has cars that can swing freely and turn upside down whether the elongated framework is stopped or revolving. Other variations on the Ferris wheel enable the whole wheel to be lowered to a horizontal position so that all cars can be unloaded and loaded simultaneously. While the cars on one of its spiderlike arms are being loaded, the rest of the cars on the Sky Whirl can continue to rotate high above Six Flags Great America theme park in Gurnee, Illinois, just north of Chicago. From the Sky Whirl, passengers today can almost make out, forty miles to the south, beyond the skyscrapers descendant from the Eiffel Tower, the phantom of the original Ferris wheel still turning in the Midway Plaisance.

HOOVER DAM

Everything seems to move slowly at Hoover Dam. Long lines of cars, buses, and trucks in low gear wind along the two-lane road down one steep side of Black Canyon and up the other. The lines stop frequently as carloads of tourists crane their necks at the grandeur of the site, and drivers search for a parking space among the tiers of overlooks. For all the congestion, few motorists seem impatient or interested in searching maps for alternate routes. Everyone driving in this vicinity must know that U.S. Highway 95, which arcs along the crest of the dam, is the only road across the Colorado River between Davis Dam—more than 50 miles to the south, near the California state line—and Navajo Bridge, the steel arch that crosses at Marble Canyon, Arizona, near the Utah border, more than 150 miles east, beyond Lake Mead and the Grand Canyon.

Pedestrians move about the top of the dam at a snail's pace in the desert heat, crisscrossing the crest over which, by design, no water has ever flowed. Several thousand tourists may visit Hoover Dam on a summer day, most of them standing in lines to board the large elevator that takes them, twenty at a time, 600 feet down into the inner workings of the dam. Many come from Las Vegas, 35 miles to the northwest. Even there, amid all the noise and neon, where casinos compete for tourist dollars, a visit to Hoover Dam is hawked as one of the area's must-do things. The pitch may be engineering achievement as awesome entertainment, but few who visit the works at Black Canyon seem disappointed by the decided calmness and harmony of it all.

While waiting to board the elevator, visitors have plenty of time to look downstream at the tamed Colorado flowing between majestic cliffs toward Lake Mohave, past Davis and other dams into Mexico and the Gulf of California—or at least that river water not shunted off into the Colorado River Aqueduct or the All-American Canal to benefit Southern California. If the ultimate fate of the water does not interest these tourists, the nonvertiginous can look from the parapets down at a sharp angle from the crest of Hoover Dam to its toe and the rectangular roofs of the buildings they eventually will enter to view the power plant. Or, in the tourist line, the conversation may turn to the large clocks near the tops of two of the four intake towers that dominate the upstream view across the dam. The clocks are labeled "Nevada Time" and "Arizona Time," and in the summer of 1993 they read the identical hour, either Pacific Daylight Time or Mountain Standard Time, as clocks were not turned forward here in spring on the Arizona side. However, this apparent gesture of concord belies the true history of the dam.

Just as the Mississippi River, which normally provides so much benefit to the communities that border it, has wreaked havoc in the Midwest, so the Colorado River used to be alternately a blessing and a bane for the Southwest. Although the great river that over millennia carved out the Grand Canyon promised to be the source of much-needed water in rich but arid regions such as Arizona and Southern California, too often it flooded low-lying lands in the spring and early summer and then in late summer and early fall slowed to a trickle, not worth diverting. If crops, cattle, and Californians were not awash, they were thirsty.

Two promising regions with rich alluvial soil were the Colorado Desert and the below-sea-level Salton Sink. Near the turn of the century they were renamed Imperial Valley by land developers who promised through their California Development Company (CDC) to supply enough water from the Colorado to make the otherwise arid land attractive. For a few years the scheme worked and the region was booming, but soon the irrigation canal silted up and, in the face of lawsuits from landowners, a new way to divert the Colorado's water had to be found quickly. A hastily engineered scheme—one that also blunted the effect of the newly formed U.S. Reclamation Service's charge that the CDC had monopolized the water supply—brought water up from Mexico. That worked fine at first, but in 1905 so much water came down the Colorado in spring and fall floods that the river changed its course and flowed into the Salton Sink, which then became the inland Salton Sea. Lost crops, lost topsoil, and a lost irrigation system amounting to millions of dollars presented a disastrous prospect for Imperial Valley. It was to be two years before the Colorado was put back on course, but the root problems associated with both exploitation of and protection from the river remained.

The great amount of silt carried southward by the Colorado kept raising the river's channel, and so required constant maintenance of the protective levees and other components of the irrigation system, much of which was located in Mexico. Problems with getting work crews back and forth across the border eventually led to support for a new canal, located entirely on American soil. It was in such a climate that a young lawyer named Phil Swing, supported by Southern California water interests, was sent in 1917 to represent them in the U.S. Congress and to promote the idea of an "All-American Canal."

Swing's effectiveness led quickly to the introduction of legislation, but it was defeated largely because of the opposition of Arthur Powell Davis, a forty-year veteran of government service who then knew about as much about the Colorado River Basin as anyone. Davis, nephew of Colorado canyons explorer John Wesley Powell, had served as chief hydrographer when a Panama Canal route was under investigation and as an engineer with the U.S. Reclamation Service since its origins in 1902. The driving force behind the design and construction of many dams and irrigation canals, he was director of the Reclamation Service when the All-American Canal bill came before the Congress. He argued successfully for its defeat in 1919 so that a more comprehensive and long-range plan for the Colorado might be explored first.

Years earlier, as a supervising engineer in the Reclamation Service, Davis had thought about a grand plan for the entire drainage system of the Colorado. According to Joseph Stevens, who subtitled his telling of the story of Hoover Dam "an American Adventure," Davis's scheme was to be "an undertaking to rival or even surpass in scale and importance the construction of the Panama Canal." Congressman Swing went further and added the Pyramids, the Great Wall of China, and Solomon's Temple to the list of great feats of engineering that were less complicated than what came to be known as the Boulder Dam Project. Congress agreed that the great problem of the Colorado Basin should be studied by the Interior Department, and its secretary, Albert Fall, assigned the task to Davis's organization. The Fall-Davis Report, issued in 1922, "contained an exhaustive hydrological and geological profile of the Colorado River and its canyons," but most attention was drawn to its recommendation that the government erect "at or near Boulder Canyon" a large dam, which would generate power to repay in time the construction expense.

Seven states would be affected by the larger plan—Arizona, California, Colorado, New Mexico, Nevada, Utah, and Wyoming—but they would first have to reach an agreement about their respective claims to water. Conferences were held, with the federal government represented by the secretary of commerce, Herbert Hoover, whom Swing claimed to have had a part in suggesting as a "neutral" member of the Colorado River Commission. It was the engineer-humanitarian Hoover who evidently broke an impasse over state-by-state allocations by proposing the establishment of Upper and Lower Colorado River basins, and this led all but one of the states to agreement. According to Hoover's account, "a blunderbuss of a governor in Arizona, who knew nothing of engineering, bellowed that it would 'rob Arizona of its birthright.' " After an amendment required ratification by only six of the seven affected states, the Colorado River Compact was accomplished late in 1922.

The Boulder Canyon Project Act was introduced in 1923 by Congressman Swing and California senator Hiram Johnson, and it became the focus of bitter debate inside and outside of Washington. The publisher of the *Los Angeles Times,* Harry Chandler, was concerned about future irrigation for the almost 1 million acres he owned just south of the Imperial Valley, in Mexico. On the other hand, William Randolph Hearst of San Francisco, Chandler's California newspaper rival, favored the bill. The saga of the Swing-Johnson legislation's debate through several sessions of Congress has been written about in detail, mostly from Swing's perspective, by Beverley Moeller. The legislative struggle finally came to an end when President Calvin Coolidge signed the bill into law in December 1928.

Even before the first Swing-Johnson bill was introduced, the Reclamation Service had begun detailed explorations of possible dam sites. When the Fall-Davis report was written, the choices had been narrowed down to five in Boulder Canyon and two about twenty miles downstream in Black Canyon. Boulder Canyon's foundation was already known to be granite, a preferred rock, whereas Black Canyon's was volcanic tuff (a compacted ash), so Davis used the language "at or near Boulder Canyon" in the report. Further investigation, however, revealed that the lower site at Black Canyon was indeed the best of the lot. Among other reasons, there was less jointing and faulting, less silt and debris to be removed, easier prospects for tunneling, and a narrower gorge that equated to a need for less concrete. Furthermore, beds of sand and gravel for use in the concrete were located nearby, the potential reservoir was larger, and nearby Las Vegas provided comparatively easy access to the canyon.

In addition to a site for the dam, the details of the design itself had to be specified. As with all engineering structures, judgment was employed to arrive at initial alternative geometries, which were then subjected to successively more refined degrees of analysis until a final design emerged. About thirty geometries were investigated at the Denver office of the Bureau of Reclamation, as the service had come to be renamed, and its engineers subjected the hypothesized dams to analyses of stresses, including those that would result from the cooling and contraction of the concrete as it aged. As was customary in the days before digital computers, models (rubber and plaster, in this case) were employed to guide and check theory and hand calculations. Initial specifications called for stresses no higher than 30 tons per square foot anywhere in the dam. In the end this proved to be difficult to meet, and stresses up to 40 tons per square foot were allowed in the final. design. This is equivalent to about 550 pounds per square inch, which is well below the compressive strength of even common concrete, thus providing a considerable factor of safety against the possibility that the dam would by crushed under its own weight or under the pressure of water it had to resist.

Although similar in vertical section to a gravity dam, one whose sheer weight suffices to prevent it from being tipped over or pushed downstream by the water, Hoover Dam acts principally as an arch dam, transferring the pressure of the water behind it to the walls of the canyon, which act as abutments. The great height of the dam, about 725 feet above bedrock, and the consequent weight of the concrete, require its transverse profile to spread like a gravity dam from 45 feet at the crest to 660 feet at the base. The structural integrity of the dam was a matter of some debate when its plans were first revealed by Elwood Mead, then commissioner of reclamation, in a 1930 article in *Civil Engineering*.

Mead outlined succinctly some "extraordinary problems met in design" in a paragraph that showed a sensitivity to scale effects and design philosophy that were essential to producing a successful outcome:

> In designing a dam more than 700 ft. in height, stress factors become very important, which in the design of dams of nominal size are comparatively insignificant. Possible errors in basic design assumptions must be carefully studied and checked; the physical properties and volumetric changes of so great a mass of concrete must be carefully determined; primary stresses caused by the weight of the materials and the horizontal water pressure must be accurately calculated, as well as secondary stresses due to all possible causes.

Mead did not elaborate on such technical matters, however, and soon an article by M. H. Gerry, Jr., a consulting engineer from San Francisco, appeared in *Civil Engineering* challenging the safety and stability of the dam. Letters challenging the challenger followed, and about a year after Commissioner Mead's article, Harald M. Westergaard, a structural-engineering professor at the University of Illinois and consultant to the bureau, published "Safety of Hoover Dam," in which he discounted Gerry's misinterpretation of structural principles and declared, "It is the business of the structural engineer to imagine each undesirable thing that might happen to the structure and provide against that." Westergaard and the bureau engineers had felt they had done just that before Mead transmitted to the secretary of the interior specifications and drawings for the dam, power plant, and appurtenant works. These were approved in late 1930, and construction bids were invited.

In his memorandum of December 15, 1930, transmitting dam specifications, Mead reminded the secretary of the interior that the Depression had created very great "pressure for action on this matter, as a means of furnishing employment and encouraging a revival of business." The specifications spelled out various conditions related to these economic concerns, including those that employment preference be given to ex-servicemen and citizens and that, specifically, "no Mongolian labor shall be employed." Other nontechnical conditions required that Boulder City be created twenty-three miles southeast of Las Vegas and close to the canyon as a construction camp site. Although bid specifications stated that buildings erected in Boulder City were to have "a reasonably attractive appearance and no unpainted shanties or tar paper shacks will be permitted," and even though there was much to be admired in the planning and construction of the town, sixty years later Joseph Stevens would relate many stories of shameful working conditions at Black Canyon.

Bids were due in Denver on March 4, 1931, but few construction companies had the experience or resources, including the $5-million bond, required to compete. The successful bidding scheme was to be put together by a group named for the task, Six Companies, Inc., and it comprised Utah Construction Co.; Pacific Bridge Co.; MacDonald-Kahn Construction Co.; Morrison-Knudsen Co.; J. F. Shea Co.; and an ad hoc group of three contractors (Bechtel, Kaiser, and Warren), which came to be collectively referred to as Kaiser Paving Co. (When the Warren Brothers firm encountered financial difficulties, the remaining seven financial backers comprised the "six companies," with Bechtel often being left off the list.) Each of the partner firms naturally had its own expertise, and Morrison-Knudsen's included "America's foremost dam builder," Frank T. Crowe.

A 1905 civil engineering graduate of the University of Maine, Crowe had gained cutting-edge experience in building high, concrete dams while he worked for the Reclamation Bureau. After almost twenty years in the field, he was offered and took a desk job as general superintendent of construction for the bureau, but he quit after a year to join Morrison-Knudsen so he could once again engage directly in dam building. It was Crowe who spearheaded the effort to come up with a bid figure for the Boulder Canyon Project, and he presented it to Six Companies representatives at a meeting early in February at the Engineers Club of San Francisco. When the bids were opened in Denver the next month, Six Companies' low bid of just under $49 million was within 0.05 percent of the price tag estimated by engineers at the Bureau of Reclamation. The contract remained, until World War II, the largest ever awarded by the government.

To build the dam proper, the river had first to be diverted through tunnels driven through the canyon walls. An upstream diversion dam, which had to be built between annual floods, and a downstream coffer dam would keep the construction site dry. After the main dam was completed, most of the diversion tunnels would be blocked off, but some parts would be incorporated into the system of penstocks that would feed the turbines in the hydroelectric power plant. After about two years, the river bottom had been cleared to bedrock and the first forms to receive concrete were erected. The pouring of concrete began on June 6,

Hoover Dam

1933, and continued day and night over the next two years. Three million cubic yards of concrete, from two specially built mixing plants, were distributed among cubelike cells that interlock in the completed dam. Cooling pipes embedded five feet apart throughout the concrete carried away the heat of hydration; otherwise, the dam would still be cooling down and developing cracks as the concrete contracted. The completed dam was turned over to the government on March 1, 1936, more than two years ahead of schedule, and energy began to be produced by the power plant that fall.

The dam was dedicated on September 30, 1935, by President Franklin D. Roosevelt. First to speak at the ceremonies was his secretary of the interior, Harold Ickes, who, after repeatedly referring to the structure as Boulder Dam, declared, "This great engineering achievement should not carry the name of any living man but, on the contrary, should be baptised with a designation as bold and characteristic and imagination-stirring as the dam itself." He was implying that the dam should not be named after Hoover, who was, of course, still alive. Ickes in his remarks had in fact reopened a debate over the name of the dam that went back to an earlier dedication ceremony, one that acknowledged Congress's first appropriations for the entire Boulder Canyon Project with the driving of a spike of Nevada silver for the rail line that was to connect the construction site with the Union Pacific Railroad in Las Vegas. At that ceremony, Ray Wilbur, the secretary of the interior under President Hoover, who signed the bill, had asserted to the surprise of many in attendance, "I have the honor to name this dam after a great engineer who really started this greatest project of all times, the Hoover Dam."

From the beginning, then, the name of the dam was a contentious and confusing issue. In 1939 the American Society of Civil Engineers (ASCE) adopted Hoover Dam for usage in society publications, pointing to correspondence between Ickes and Attorney General Homer Cummings. Cummings declared the name Hoover Dam to be official, because of its use in the appropriations bill and government contracts for the dam, as opposed to the collective Boulder Canyon Project, which included also the power plant and appurtenant works. In 1947 the Republican Eightieth Congress passed legislation reinstating the name Hoover Dam. Whatever its name, more than 27 million people have visited the dam over the years, and there appears to be general agreement with a plaque—placed near the center of the crest by the ASCE in 1955—declaring the dam to be one of the country's Seven Modern Civil Engineering Wonders.

THE CHANNEL TUNNEL

A common misconception about technological development is that engineers go ahead willy-nilly with whatever is physically possible at a given time and expect the rest of the world to fall into place and follow. This viewpoint did not have its origins in the counterculture movement of the 1960s, nor did it begin in reaction to the development of atomic weapons during World War II. The thought was embodied in the familiar but erroneous slogan of the 1933–34 Century of Progress Exposition in Chicago ("Science finds—industry applies—man conforms"), but it predated even that. We can find the idea—that scientific and technological advances carry us where we might not otherwise go—expressed during the nineteenth century in the thoughts of Ralph Waldo Emerson ("Things are in the saddle, / And ride mankind") and his metaphor was no doubt effective because the misconception was already a familiar one.

In fact, engineers and inventors have seldom been in a position to implement their ideas single-handedly. Technology is a social enterprise requiring the cooperation and support of lawyers, bankers, businesspeople, and other nontechnical participants. This can be true even of the smallest things, if they are to have an impact worthy of protest, for few if any inventors have the inclination to deal equally with the patent process, the chore of raising capital, and the details of marketing and sales.

For larger technological artifacts, such as the interstate highway system, the engineer cannot even begin to be effective without the direct financial support and approval of government agencies. Although perhaps more formal and explicit today than in the past, restrictive legislation, regulations, impact statements, hearings, and other forms of constraint on technology by society have always been present, if only by edict or influence. And if such can be the social obstacles to national enterprises, imagine how much more difficult it must be to bring to fruition engineering ideas of an international scope.

A fixed link between England and France is now a reality, but it emerged as a technological idea almost 250 years ago, when it was articulated by the eighteenth-century French engineer Nicolas Desmaret. This is not to say that he was the first to stand on the cliffs at Calais or Dover and imagine how a tunnel might be driven under the channel, but just as philosophers can ask if a tree falling in the forest makes a sound if there is no one around to hear it, so we can ask if an individual imagining a tunnel constitutes an engineering idea if it is offered to no one for criticism. Since engineering is a social enterprise, totally private imaginings or dreams cannot be said to qualify as engineering. And even proposing an engineering project, though obviously necessary, is a far-from-sufficient step in realizing it. James Kip Finch saw this as a theme of his remarkable history *Engineering and Western Civilization,* and he began the preface to his book with a quote to this effect from a 1927 address to engineering students at Columbia University: "It is not the technical excellence of an engineering design which alone determines its merit but rather the completeness with which it meets the economic and social needs of its day."

Although Desmaret's eighteenth-century tunnel scheme may not have had even technical excellence, it did put the matter on the table for future engineers to pick up on. In 1802 Albert Mathieu, a French mining engineer, displayed in Paris plans for a tunnel more than eighteen miles long under the channel. The tunnel was to be lighted by oil lamps, and ventilation was to be provided through chimneys emerging above the sea. An artificial island about midway across the channel was to provide a place to change horses, which were to pull coachloads of passengers and mail at a gallop between England and France. Mathieu's idea came to the attention of Napoleon Bonaparte, who was then First Consul, and he thought enough of it to discuss it with the English statesman Charles James Fox at a meeting during the Peace of Amiens. Fox embraced the idea as one the two nations could undertake cooperatively. As with many such great projects, alternative ideas were also soon proposed, including one the following year comprising a submerged tube on the seabed. With cooling relations between the countries, however, the ideas languished.

The earliest tunnel schemes were proposed before much was known about the channel bottom, but in 1833 a young French civil engineer and hydrographer named Aimé Thomé de Gamond began systematically what was to be his thirty-four-year attack on the problem. He established by soundings the profile of the channel bottom, and he proposed several crossing schemes: a prefabricated iron tube laid down from the sea surface, a masonry tube left behind a movable undersea shield, and five different bridge schemes. The "obstinate resistance of mariners" and the high costs of bridges led Thomé de Gamond to explore the hitherto unknown geological conditions under the channel proper. His crude observations, some said to have been made by his personally diving to depths of a hundred feet or more, convinced him that a tunnel could be mined beneath the channel.

Thomé de Gamond's tunnel was to be twenty-one miles long, over which would be located twelve artificial islands containing ventilation shafts. The shafts were also to contain "sea valves," so that if war were to break out between England and France, the tunnel could be flooded. Such a feature was, no doubt, to obviate the objection that a tunnel would provide an invasion route, a fear that England more than France would hold for many years to come. In 1856 Thomé de Gamond gained an audience with Napoleon III, who afterward had a scientific commission evaluate the plan. The commission was favorably impressed by the idea, but it felt that more technical information was required to make a determination. Furthermore, a preliminary agreement between the British and French governments was felt to be needed before proceeding with the technical details.

Encouraged by the commission's response, Thomé de Gamond contacted prominent British engineers and gained the support of, among others, Isambard Kingdom Brunel, who had worked on his father's subaqueous Thames Tunnel and his own two-mile-long railroad tunnel through the oolite ridge at Box. Prince Albert was also enthusiastic about the idea of a channel tunnel, and so was Queen Victoria, who suffered from seasickness and would have welcomed an alternative way to cross the channel. During Thomé de Gamond's visit to England in 1858, however, the prime minister, Lord Palmerston, spoke for the many Britons who preferred to retain their island status, when he said to the French engineer, "What! You pretend to ask us to contribute to a work the object of which is to shorten a distance which we find already too short!" In that same year, an assassination attempt on Napoleon III employing Birmingham-made bombs cooled the French interest and also renewed English fears of an invasion.

In the meantime, railroads had become well established on both the island and the Continent, leading to competing schemes to accommodate them in a channel tunnel. However, since the coal-burning and soot-spewing locomotives could barely be tolerated in the fifteen-hundred-foot-long tubes of the Britannia Bridge, it was unlikely that they would work well in a twenty-mile undersea tunnel. Another French engineer, Hector Horeau, proposed to solve this problem by having submerged iron tubes just so inclined that steam engines would not be necessary to pull the rolling stock, except to assist in hauling the cars out from where they ran out of kinetic energy. Some years later, with the successful driving of the Mount Cenis rail tunnel through more than eight miles of the French Alps and the work on the nine-mile St. Gotthard Tunnel through the Swiss Alps, along with the growing interest in international trade, came renewed interest in a subchannel tunnel through which rail traffic could move freely.

It was a British engineer, William Low, with considerable experience in Welsh coal mines and their ventilation, who came up with the scheme for driving a pair of parallel, single-track rail tunnels, connected intermittently by cross passages. The piston action of the trains in these interconnected tunnels would replenish the air and make them self-ventilating. Low's idea formed the basis of the plan eventually advanced a century later by the Channel Tunnel Study Group. By the early 1870s, Low's idea was agreed to in principle by the British and French governments, but neither would make any financial commitment. Competing private companies thus began to vie for the job, and by the mid-1870s serious surveys, including the taking of 7,700 soundings and 3,267 geological samples, confirmed that the so-called Lower Chalk stratum appeared to extend unbroken under the channel, except for some locations near each shoreline.

The Lower Chalk, also known as Cenomanian or Chalk Marl, is ideal for tunneling, because it is essentially flint-free and almost impermeable to water, especially in its lower levels, where it is mixed with clay. The properties of the Lower Chalk are in contrast with the Upper and Middle Chalk strata, which are laden with hard flint, contain many fissures, and have water present. It was Sir Edward Watkin, a railroad man, who took up Low's scheme and in 1880 had a pilot tunnel driven about a half mile into the Lower Chalk, confirming its properties. The tunnel was driven with a patented tunneling machine commonly credited to Colonel Fred Beaumont but said to have been designed by one Thomas English. The machine cleverly employed compressed air for both powering the revolving cutters and ventilating the working area, and it was the subject of much contemporary discussion. The tunnel became the site of many visits and promotional dinners, all of which encouraged support for the project, and by early 1882 it had reached nearly two-thirds of a mile.

Despite the technological success, political opposition returned. After the death of Prince Albert, Queen Victoria was less enthusiastic, and Prime Minister William E. Gladstone had reservations about connecting "Happy England" to the Continent. Others joined the chorus. "As an improvement in locomotion, and as a relief to the tender stomachs of passengers who dread sea-sickness, the design is excellent," London's *Times* editorialized, but "from a national point of view it must not the less be received with caution." Even assurances that fortifications could be constructed at the English opening and explosives could collapse the tunnel on invading armies did not assuage the xenophobia. When schemes to flood the tunnel were assured by the engineers, the editor of *The Nineteenth Century* invoked Shakespeare to defend the island:

> *This royal throne of kings, this scepter'd isle,*
> *This earth of majesty, this seat of Mars,*
> *This other Eden, demi-Paradise,*
> *This fortress built by Nature for herself*
> *Against infection and the hand of war,*
> *This happy breed of men, this little world,*
> *This precious stone set in the silver sea,*
> *Which serves it in the office of a wall*
> *Or as a moat defensive to a house,*
> *Against the envy of less happier lands. . . .*

The editor further declared, forsaking poetry for fear, that for engineers "to hang the safety of England at some most critical instant upon the correct working of a tap, or of any mechanical contrivance, is quite beyond the faith of this generation of Englishmen."

Progress on the so-called Beaumont Tunnel was stopped by the Board of Trade, which ruled that, no matter whether Sir Edward Watkin owned the land from which he accessed his tunnel, it was the Crown that controlled even the Lower Chalk from the low-water mark on the shore to the three-mile limit. (The unlined tunnel would remain in good condition for a century, thus giving confidence to twentieth-century attempts at tunneling.) Although political opposition prevailed, an aging Sir Edward continued his efforts into the mid-1890s, when he finally gave up.

The idea of a channel tunnel was revived about a decade later. Relations between England and France had improved, and in the interim electric locomotives had been introduced, thus bringing a general renewal of interest in subaqueous railroad tunnels, including under the Hudson River that would bring the Pennsylvania Railroad from New Jersey into New York.

Improvements in tunneling, such as the chore of getting rid of the spoil, had removed any remaining doubts about the technology, but political opposition remained. In an attempt to allay the fears of the English, the French proposed that the single entrance on their shore be along a great horseshoe-shaped viaduct that stretched far into the sea before curving around to join the coastal rail line. Thus, in times of war, the British fleet could control this access and shell it into inoperability if necessary.

With World War I came the airplane, and England's island status suddenly became less protective. After the war, support for a tunnel seemed finally in place, but postwar conditions were used as an argument against it. There was a brief revival of interest in the late 1920s, but that ceased because of the economic conditions of the 1930s. World War II presented concerns among some English that the Germans would tunnel secretly under the channel, but others argued that had a tunnel been in place the British could have been more effective in fighting the war on the Continent. There was little talk of the tunnel after the war, in part because rebuilding the country was more urgent and in part because the offices and records of the Channel Tunnel Company were destroyed in the Blitz.

By the mid-1950s, when atomic bombs and jet aircraft had put military arguments in a new light and had made the world smaller, there began to be international interest in financing a tunnel. The idea attracted promotors of macroprojects, such as the international lawyers Frank Davidson and Cyril Means, Jr., who had read about the nineteenth-century attempts to tunnel under the channel. They sought to involve the Suez Canal Company, for its experience with large projects, with the languishing British and French tunnel companies, and a new Channel Tunnel Study Group was formed in 1957 to revisit the problem. Modern electronic geophysical equipment was employed, along with deep rock borings, to recharacterize the site, and new economic projections of use and revenue were made. Furthermore, various tunnel and bridge alternatives, most of which followed the nineteenth-century routes and schemes, were investigated for their technical feasibility and cost. In 1960 the study group concluded that the best solution was a rail tunnel in which motor vehicles would be carried on railcars, the scheme that was finally to be successful.

At that time, most vocal opposition to the study dealt with financial issues. Considering alternative means to cross the channel encouraged protests from competing commercial interests, such as ferry services and supporters of the developing hovercraft concept. The original proposal of the study group was that private interests provide only the tunnel proper, with the British and French governments adding approach and support facilities and rolling stock. Amid opposition, the group later agreed to finance with private funds the terminals and approaches, thus minimizing commitments of public funds, but the enterprise languished.

Even with no direct financial interests, government approvals for a channel crossing remained necessary, however, and these came finally in the mid-1970s in the form of a Franco-British treaty. The Channel Tunnel Company and the Société Française du Tunnel Sous la Manche were to coordinate construction with the British and French railroads, and all were to share in the revenues for the first fifty years of operation, after which ownership would go to the respective governments. Preliminary contracts were let, and tunneling work started once again in anticipation of the final ratification of the treaty. In 1975, however, just before approval, the British pulled out of the project, ostensibly in part over uncertainties regarding the modernization of the rail line between London and the coast.

There were various attempts to save the tunnel project, and by the early 1980s a new consortium—comprising British Rail, the Société Nationale de Chemin de Fer Français, and the private European Channel Tunnel Group—was formed. This provided renewed interest in the project, and when Prime Minister Margaret Thatcher and President François Mitterrand met for the first time in London, in 1981, they announced formation of the Joint Technical Commission to discuss the matter further. At a subsequent summit meeting an open invitation to promoters was issued, with proposals due on October 31, 1985.

Ten schemes were submitted, and four were considered seriously. These were the Eurobridge, a multilevel roadway enclosed completely inside an ellipsoidal tube; the Euroroute, an assemblage of tunnels, bridges, and artificial islands; the Channel Expressway, comprising four separate tunnels; and the Eurotunnel, which was the accepted design. This rail-tunnel idea, which would include a shuttle service for motor vehicles, was proposed jointly by the Channel Tunnel Group and France Manche. It is essentially the same as the 1950s concept and is remarkably like the plan proposed by William Low, begun by Sir Edward Watkin, but stopped by the British Board of Trade more than a century earlier.

With the project and a fifty-five-year concession awarded to Eurotunnel, as the Anglo-French promoters came collectively to be known, preliminary site work could begin in 1987, with service expected to begin in 1993. There were, of course, delays, but considering the magnitude and history of the project, it held remarkably close to schedule—and commercial traffic was flowing by mid-1994. Unfortunately, after the first year or so of service, revenue from the traffic was not providing enough cash to pay off the interest on the debt, and Eurotunnel was in financial trouble.

The "Chunnel," as it has become familiarly known, will no doubt continue to carry traffic between England and France, and it will stand as one of the late twentieth century's most remarkable technological accomplishments. However, the technology, albeit more primitive, was in place during earlier attempts, and an old boring machine remains buried in a blind, abandoned Beaumont tunnel to this day as a monument to that fact. The execution and success of projects such as the Channel Tunnel are clearly as much about economics and politics and desire as they are about engineering and technology.

THE PETRONAS TOWERS

For two decades, the 110-story Sears Tower stood as the tallest building in the world. At 1,454 feet above ground level—almost as tall as a string of five football fields would be long—the bundle of nine steel tubes standing just outside Chicago's Loop could be said to have cast a shadow over all other skyscrapers since its completion in 1974. New York's World Trade Center, the 1,368- and 1,362-foot-tall twin towers completed only a year earlier, held the record as tallest buildings for only a brief time. Before that, the Empire State Building, standing at 1,250 feet even without its broadcasting towers, which, like those of the Sears Tower, do not count as part of the building proper, held the world height record for more than four decades. Finished in 1931, the Empire State then surpassed the one-year-old Chrysler Building, which at 1,046 feet had been the first to break the magical 1,000-foot mark. Before then, the Woolworth Building, a 792-foot Gothic cathedral of commerce paid for in cash by the profits from its namesake's chain of five-and-ten-cent stores, was the world's tallest building for almost two decades.

Skyscrapers—so named since the 1880s, when Chicago's 100-foot-tall buildings were marvels of contemporary structural engineering—seem to have sprouted up in temporal and spatial clusters, with Chicago and New York proving to be especially hospitable to the form and its financing. Throughout most of the twentieth century, the skyscraper was considered a peculiarly

American genre, growing with the economy and optimism of cities such as Atlanta, Houston, Los Angeles, and Seattle. In the last decade of the century, however, the frontier of the skyscraper has moved across the Pacific Ocean to the Far East. Today most of the tallest buildings in the world are being proposed for locations such as Japan, Taiwan, Hong Kong, and mainland China. And they are not only being proposed, they are also being built, with a new tallest building in the world having topped out in 1996 at 1,482 feet in Kuala Lumpur, Malaysia.

The Federation of Malaysia is a Southeast Asian country of about 20 million people, the vast majority of whom live in West Malaysia, located just above the equator, on the Malay Peninsula, between Thailand to the north and Singapore to the south. (As established in 1963, the federation included Singapore, but it seceded in 1965.) The two states of East Malaysia are situated to the east, across the South China Sea, on Borneo, the third-largest island in the world. Kuala Lumpur, the capital of Malaysia, is on a modern north–south expressway that puts Singapore within a five-hour drive. The city's Subang Airport has direct flights throughout the world and hourly flights that depart for and arrive from Singapore, providing a bird's-eye view of lush golf courses among the rubber trees—a sign of the changes that have been taking place in Malaysia, which has been described as "predominantly a Malay Islamic state with strong Chinese and Indian influences."

In the 1980s, the Malaysian economy was based on commodities such as palm oil, rubber, and tin. However, with the rise of a government headed by the strong-willed and powerful Prime Minister Mahathir bin Mohammad, by training a medical doctor, Malaysia began to assert itself as a country on the move. The prime minister is said to have "reinterpreted Islam as allowing the pursuit of wealth and technical knowledge," and the government's official objective came to be articulated, in Bahasa Malaysian, the country's national language, as Wawasan 2020—or Vision 2020 in English, which is the country's accepted language of commerce and industry. The vision has Malaysia transformed into a fully developed industrialized nation by the year 2020, with manufacturing and construction to be the main driving forces of the country's economy. The Kuala Lumpur City Centre (KLCC) project is helping in a big way to make the Malaysian vision a reality.

Kuala Lumpur City Centre is a hundred-acre development on the site of a former racetrack, the Selangor Turf Club, and is one of the largest real-estate development projects in the world. As the result of an international competition held in 1990, a master plan—"an ideal city-within-a-city"—was defined by the U.S. firm of Klages, Carter, Vail & Associates of Cosa Mesa, California. It comprises a fifty-acre park, which will include a lake, much of which will be accessible to the public, and a complex of twenty or so surrounding buildings that will contain office space, apartments, hotel rooms, recreational facilities, restaurants, shops, banks, a convention center, a civic center, a mosque, and a plant to provide chilled water for cooling all these buildings in the tropical climate. The first phase of the $2 billion project includes the pair of buildings known as the Petronas Towers, themselves costing approximately $800 million, most of which was to be provided by Petroleum Nasional Berhad, Malaysia's national oil company and the source of the towers' name, and the government. As they have risen to become the world's tallest buildings, these striking towers have already come to be among Kuala Lumpur's most significant landmarks.

Back in 1991, with the master plan in hand, a separate international design competition was held to determine exactly what kind of structure would provide a significant focal point and monumental entrance to Kuala Lumpur City Centre. The competition was won by the architects Cesar Pelli & Associates, of New Haven, Connecticut. According to Pelli, the client wanted a Malaysian image but could not say exactly what that meant. Existing tall buildings in Kuala Lumpur were of the boxy international style. With no indigenous structural models to inspire him, Pelli looked to Islamic art and adopted a multipointed star pattern as a footprint for his building design. His early scheme called for a twelve-pointed star layout, but this was changed to a modified

The Petronas Towers

eight-pointed floor plan with intermediate arcs when the prime minister observed that the former geometry was more Arabic than Malaysian.

The final design of skyscrapers, especially those that are to be the tallest in the world, do not evolve wholly from an architect's drawings. How a structure will stand against the forces of nature—whether the ground motion in an earthquake zone, the wind at hundreds of feet above the ground, or the heat of the sun beating down on the ground from above—requires the insight and calculation of a structural engineer. (The exposed structure of Chicago's John Hancock Tower, for example, was the result of a collaboration between the architect Bruce Graham and the structural engineer Fazlur Khan, both of the firm of Skidmore, Owings & Merrill.) Pelli wished the "cosmic pillars" in Kuala Lumpur to be joined by a skybridge to form a welcoming portal to KLCC and to have as few structural columns as possible blocking the view outward from the office floors. Such features were easy to render on the drawing board but no simple task to realize in concrete and steel. For the Kuala Lumpur project, Pelli sought the structural expertise to accomplish such objectives in the international engineering design firm of Thornton-Tomasetti Engineers, based in New York City. Charles H. Thornton, chairman and principal in the firm, had long wanted to design the tallest building in the world, and had indeed designed with Pelli a 125-story Miglin-Beitler Building for Chicago that for financial reasons did not come to fruition. Thornton was thus well prepared to work again with Pelli on the Petronas Towers, equivalent to 95 stories in height.

The challenges associated with designing and building a skyscraper begin in the ground. If the foundation is not firm, the building will be susceptible to settling, which, in the worst case, can lead to tilting and collapse. Underground conditions are not often fully known, however, until extensive, albeit usually still only sampling, exploration takes place, and this may not proceed until the design is defined enough for engineers to set the locations and types of tests needed. As it turned out, the Kuala

Lumpur master plan had sited the showcase buildings over an underground cliff. To found piles in the rock, which sloped very steeply and contained caverns, would have meant that every pile location would have had to be surveyed before proceeding. Thus the location of the towers was moved about 200 feet to the southeast, where the generally fissured limestone was sufficiently deep so that all piles could terminate in the ground above the hard rock and thereby ensure a more uniform foundation. This alluvial ground, known as Kenny Hill, was in fact once itself hard rock, but after being exposed to prolonged weathering under tropical conditions, it became soft for hundreds of feet below the surface. What is thus considered soil by most engineers is still considered rock by geologists. Whatever the ground's composition, however, the challenging foundation conditions necessitated drilling some piles almost four hundred feet deep, more than three times the depth of the foundation beneath the Sears Tower. Even then, the foundations of the new buildings are expected to settle as much as three inches under the weight of the completed structure.

Designing the superstructure of a building presents another set of challenges. Among the first decisions facing structural consultant Thornton and his associates at Ranhill Bersekutu in Malaysia was the choice between steel and concrete. Although the tallest skyscrapers are steel structures, that material was not readily available in Malaysia, where prohibitively high tariffs on imported steel make concrete the construction material of choice. Furthermore, steel buildings tend to be more flexible than concrete and sometimes have to be fitted with mechanical devices known as tuned mass dampers to ameliorate the effects of vibrations induced by the wind. Concrete structures, on the other hand, while they tend to be stiffer and have qualities that damp out quickly any vibrations that do begin, often are bulkier-looking than steel. Since the architect wished the Petronas towers to be slender-looking and have columns spaced rather far apart, conventional concrete columns would have been too aesthetically broad and structurally heavy. To overcome this objection, an extremely high-strength concrete was developed, with bearing capacities as many as three times that conventionally used in Malaysia—or elsewhere, for that matter. Special concrete mixes, using local materials, were developed to produce compressive strengths as high as 10,000 pounds per square inch, with quality control provided by state-of-the-art computerized systems.

Being made of high-strength concrete, the columns around the periphery of the towers could be smaller in diameter and lower in mass, thus reducing their deadweight. Still, at the base of the towers, the columns are nearly eight feet in diameter. The towers are not purely concrete structures, however, and the floor beams spanning between each building's core and ring of columns are made of steel. This was done to speed construction, minimize the floor height, and better accommodate such mechanical equipment as cooling ducts. The tapering at the top of each building demanded some especially tricky structural engineering, and its geometry necessitated installation of a wide variety of different-size glass panels. The record height of the towers is achieved through the pinnacles at their tops, which are part of the basic architecture and structure proper, unlike the broadcast antennas erected after the fact on such buildings as the Empire State Building and the Sears Tower. (Once the pinnacles of the Petronas Towers were in place, the official arbiter of skyscraper records, the Council on Tall Buildings and Urban Habitat, confirmed that the Sears Tower had indeed been surpassed as the tallest building on earth.) The pinnacles, with designs based on minarets rather than Gothic cathedral spires, were erected by jacking them skyward from within the uppermost part of the hull of the towers. Detailed studies included many options, and the final pinnacle design is a scaled-up version of one originally proposed to provide a graceful tower top and, coincidentally, reach a record-breaking height. Thornton, an avid sailor, likens each pinnacle's structural support to that of the mast of a sailboat.

Among what makes extremely tall buildings viable investments is the amount of usable, rentable, or salable floor space they contain relative to their height. As buildings grow taller, more and more of their volume must be devoted to elevators to transport the tens of thousands of occupants up and down. In the Petronas towers, the usable floor space was increased considerably by the addition of smaller forty-four-story structures, referred to as "bustles," to the towers. With the bustles, which are topped by prayer rooms for the Muslim occupants who are called to prayer twice each working day, each tower has about two million square feet of office space. The internal transportation system that will move people vertically in the towers will include double-deck express elevators to optimize the use of the shafts. Passengers will transfer to and from double-deck local elevators in sky lobbies about halfway up the buildings, on the forty-first and forty-second floors.

The towers are also connected to each other at the level of the sky lobbies by a skybridge, a 190-foot-long steel walkway that not only facilitates movement between the top of one tower and that of the other but also serves as an alternate escape route in the event of a fire or other emergency, such as occurred when a terrorist bombing forced the evacuation of one of the towers of the World Trade Center. But the skybridge's design presented additional and unusual structural problems. Since the two towers can sway in the wind both in phase and out of phase, as well as twist in independent directions, the skybridge could not be attached rigidly between the vertical structures. Thus special bearing connections had to be devised to allow for as much as 12 inches of horizontal movement each way at each end, as well as the twisting. Because such a long skybridge would have to be of very heavy construction if it were not to sag in the middle, a set of slender steel legs was designed to angle up from supports on each tower about 160 feet below to the center of the skybridge. To prevent such slender struts from vibrating excessively in the wind, thus presenting the potential for the growth of fatigue cracks, tuned mass dampers were designed and installed inside the legs. (Each mastlike pinnacle also has a damper in the form of an energy-absorbing, rubber-sheathed chain.)

The Petronas Towers and all the buildings planned for the Kuala Lumpur City Centre are what are known as intelligent buildings, employing automatic controls and communications systems to minimize energy consumption and maximize the comfort of occupants and the convenience of use. The concept of an intelligent building dates from the 1980s, when costs associated with installing and retrofitting environmental and communications systems were escalating. There also came to be recognized clear advantages in incorporating networking capabilities into a new building, rather than providing tenants with a structural shell that they themselves have to wire. In the Petronas Towers, each floor or pair of floors has its own local area network for air conditioning and lighting, as well as a general-purpose controller for unspecified future use.

Although the record-breaking height of the Petronas Towers is their most immediately visible and talked-about feature, that notoriety is likely to be their most short-lived, since taller buildings are being planned for Shanghai, China, and Melbourne, Australia. Among the more long-range benefits of the Petronas Towers to the Malaysian economy is the considerable amount of technology transfer that accompanied their design and construction, with the direction and support of the prime minister. For example, the development of the high-strength concrete used in the towers in four years doubled the strength of concrete produced in Malaysia. This meant that buildings could now be built in less time and for less money.

Another by-product of the Petronas Towers project was the establishment of new local industries. The towers required about 1.5 million square feet of stainless steel cladding and glass, in the guise of 32,000 windows, to form a so-called curtain wall. This was the largest such job that Harmon Contract, the Minnesota-based participant in the cladding contract consortium, had ever tackled, and to win the job, Harmon had prepared an eight-hundred-page bid to meet the Malaysian expectation that proposals be educational as well as business instruments. Furthermore, as a condition of receiving such a large contract, Harmon was required to set up shops in Malaysia to fabricate the components of the curtain wall and thereby introduce a new industry to the country.

The experience gained by local engineers and contractors in designing and building the world's tallest buildings has prepared them well to complete the rest of Kuala Lumpur City Centre by the year 2020. By demanding local participation in the project, the Malaysian government, led by its strong-willed prime minister, ensured that the Petronas Towers project would leave a legacy that will serve the local economy and society long after taller buildings are erected elsewhere in the world.

Acknowledgments
and Bibliography

IMAGES OF AN ENGINEER

After this column appeared in *American Scientist* (for July–August 1991), several readers helped clarify details about Steinmetz's physical condition and the altered photograph. Among these was Hartmann H. R. Friederici, M.D., who noted that Steinmetz was no dwarf, for his bones were of normal length. However, his spine was twisted and angled in a pattern known medically as kyphoscoliosis, a not-uncommon deformity in earlier centuries in Central Europe that affected such notables as the Hunchback of Notre Dame. The likely cause, tuberculosis of the spine, was a common disease in those times, and Steinmetz might have been as tall as Einstein had the deformity not resulted. Dr. Friederici also further noted that the likely cause of Steinmetz's relatively early death was the burden placed on the lungs by their being unable to expand properly, and on the heart, probably resulting in right-sided heart failure.

Sidney Metzger, whose letter to the editor appeared in the November–December 1991 issue of *American Scientist,* provided details about and a copy of the group photo from which Einstein and Steinmetz were extracted. Metzger also pointed out that Steinmetz enjoyed trick photography. In a letter to me, Israel Katz described the collection of Steinmetz's own photos that Katz had seen when as a young GE test engineer he was renting a room from relatives of Steinmetz's housekeeper and cook, who had preserved the photos.

Dos Passos, John. *U.S.A.* New York: Modern Library, 1939.

Hammond, John Winthrop. *Charles Proteus Steinmetz: A Biography*. New York: Century, 1924.

Jordan, John M. " 'Society Improved the Way You Can Improve a Dynamo': Charles P. Steinmetz and the Politics of Efficiency." *Technology and Culture* 30 (January 1989): 57–82.

Kline, Ronald R. *Steinmetz: Engineer and Socialist*. Baltimore: Johns Hopkins University Press, 1992.

Kline, Ronald. "Manufacturing Legend: Charles Proteus Steinmetz as Modern Jove." *Engineering: Cornell Quarterly*, Autumn 1989, pp. 49–54. See also letter, ibid., Winter 1990, pp. 59–60.

LaFollette, Marcel C. *Making Science Our Own: Public Images of Science, 1910–1955*. Chicago: University of Chicago Press, 1990.

Leonard, Jonathan Norton. *Loki: The Life of Charles Proteus Steinmetz*. New York: Doubleday, Doran, 1930.

Nye, David E. *Image Worlds: Corporate Identities at General Electric, 1890–1930*. Cambridge, Mass.: MIT Press, 1985.

Wise, George. *Willis R. Whitney, General Electric, and the Origins of U.S. Industrial Research*. New York: Columbia University Press, 1985.

ALFRED NOBEL'S PRIZES

A shorter version of this essay was first published in *Issues in Science and Technology* (Fall 1987). The English translation of Alfred Nobel's will that is quoted is one that appears to have some official status, it being featured in a booklet issued by the Nobel Foundation, and it also being reproduced without comment by Elisabeth Crawford in her seminal study of the Nobel archives relating to the origins of the science prizes as we know them today.

Bernhard, C. C., et al., eds. *Science and Technology in the Time of Alfred Nobel*. Oxford: Pergamon Press, 1982.

Crawford, Elisabeth. *The Beginnings of the Nobel Institution: The Science Prizes, 1901–1915*. Cambridge and Paris: Cambridge University Press and Éditions de la Maison des Sciences de l'Homme, 1984.

Heilbron, J. L. *"Fin-de-Siècle* Physics." In *Science and Technology in the Time of Alfred Nobel,* ed. C. C. Bernhard et al. Oxford: Oxford University Press, 1982.

Petroski, Henry. "The Draper Prize." *American Scientist,* March–April 1994, pp. 114–17.

Stahle, Nils K. *Alfred Nobel and the Nobel Prizes.* Stockholm: The Nobel Foundation, 1978.

Watson, J. G. *A Short History of the Institution of Civil Engineers.* London: Thomas Telford, 1982.

Wilhelm, Peter. *The Nobel Prize.* London: Springwood Books, 1983.

HENRY MARTYN ROBERT

This essay first appeared in *American Scientist* for March–April 1996. Catherine Petroski provided me with her research on Robert from the collections of the Library of Congress.

Mehren, E. J. "Henry Martyn Robert." *Engineering News-Record,* April 22, 1920, pp. 798–802.

Robert, Henry M. *Robert's Rules of Order, Revised,* 75th anniversary (6th) ed. Chicago: Scott, Foresman, 1951.

———. *The Scott, Foresman Robert's Rules of Order Newly Revised,* 9th ed., by Sarah Corbin Robert. Clenview, Ill.: Scott, Foresman, 1990.

Smedley, Ralph C. *The Great Peacemaker.* Santa Ana and Los Angeles, Calif.: Toastmasters International and Borden, 1955.

JAMES NASMYTH

This essay appeared, in a somewhat shortened form, in *Mechanical Engineering* for February 1990.

Emmerson, George S. *Engineering Education: A Social History.* Newton Abbot, Devon, Eng.: David & Charles, 1973.

Ferguson, Eugene S. "The Mind's Eye: Nonverbal Thought in Technology." *Science* 197 (August 26, 1977): 827–36.

Nasmyth, James. *James Nasmyth, Engineer: An Autobiography,* ed. Samuel Smiles. London: John Murray, 1885.

Newton, Isaac. *The Correspondence of Isaac Newton. Vol. I: 1661–1675,* ed. H. W. Turnbull. Cambridge: Cambridge University Press, 1959.

Rolt, L. T. C. *Victorian Engineering.* Hammondsworth, Middlesex, Eng.: Penguin, 1970.

ON THE BACKS OF ENVELOPES

After this essay appeared in *American Scientist* (January–February 1991), one reader, Dr. Gilbert J. Sloan, of Wilmington, Delaware, wrote that creativity of the kind John Stevens demonstrated by drawing on the back of his wife expresses itself similarly in different fields. In support of his point Dr. Sloan provided a quote from Goethe's *Roman Elegies,* in which the poet describes days and nights with his beloved:

> *When sleep overcomes her, I lie and think deeply.*
> *Often, I even wrote poems in her embrace*
> *And with fingering hand, counted*
> *The hexameter's measure on her back . . .*

Abbott, Philip G. "Introduction to Steel Supplement." *Civil Engineering* (London), November–December 1985, p. 5.

Allen, Oliver E. "The First Family of Inventors." *American Heritage of Invention & Technology,* Fall 1987, pp. 50–58.

Bartlett, John. *Familiar Quotations,* 14th ed. Boston: Little, Brown, 1980.

Baynes, Ken, and Francis Pugh. *The Art of the Engineer.* Woodstock, N.Y.: Overlook Press, 1981.

Billings, Henry. *Bridges.* New York: Viking, 1966.

Furhmann, Henry. "Why Things Fall Down." *Columbia,* April 1988, pp. 31–36.

GOOD DRAWINGS AND BAD DREAMS

This essay first appeared in *American Scientist* for March–April 1991.

Baynes, Ken, and Francis Pugh. *The Art of the Engineer.* Woodstock, N.Y.: Overlook Press, 1981.

Ferguson, Eugene S. *Engineering and the Mind's Eye.* Cambridge, Mass.: MIT Press, 1992.

Friedel, Robert, and Paul Israel. *Edison's Electric Light: Biography of an Invention.* New Brunswick, N.J.: Rutgers University Press, 1986.

Gordon, J. E. *Structures: Or, Why Things Don't Fall Down.* New York: Da Capo Press, 1978.

Haldane, J. W. C. *Life as an Engineer: Its Lights, Shades and Prospects.* London, 1905.

Harris, P. R. *The Reading Room.* London: The British Library, 1979.

Hoover, Herbert. *Memoirs: Years of Adventure, 1874–1920.* New York: Macmillan, 1952.

Leonhardt, Fritz. *Bridges: Aesthetics and Design.* Cambridge, Mass.: MIT Press, 1984.

Rolt, L. T. C. *The Railway Revolution: George and Robert Stephenson.* New York: St. Martin's Press, 1962.

Tichi, Cecelia. *Shifting Gears: Technology, Literature, Culture in Modernist America.* Chapel Hill: University of North Carolina Press, 1987.

van der Zee, John. *The Gate: The True Story of the Design and Construction of the Golden Gate Bridge.* New York: Simon & Schuster, 1986.

FAILED PROMISES

This *American Scientist* column (January–February 1994) was prompted by the article by Leveson and Turner referenced below and brought to my attention by Mike May and Rosalind Reid.

Anonymous. "Ship Collisions Are Scheduled for Testing Validity of Computer Modelling." *Finite Element News,* April 1992, p. 40.

Casey, Steven. *Set Phasers on Stun: And Other True Tales of Design, Technology, and Human Error.* Santa Barbara, Calif.: Aegean Publishing, 1993.

Ferguson, Eugene S. *Engineering and the Mind's Eye.* Cambridge, Mass.: MIT Press, 1992.

Jakobsen, B. "The Loss of the Sleipner A Platform." *Proceedings of the Second International Offshore and Polar Engineering Conference.* San Francisco, June 1992, pp. 14–19.

Leveson, Nancy G., and Clark S. Turner. "An Investigation of the Therac-25 Accidents." *Computer,* July 1993, pp. 18–41.

Neumann, Peter G. *Computer-Related Risks.* New York and Reading, Mass.: ACM Press and Addison-Wesley, 1995.

Peterson, Ivars. *Fatal Defect: Chasing Killer Computer Bugs.* New York: Times Books, 1995.

Wiener, Lauren Ruth. *Digital Woes: Why We Should Not Depend On Software.* Reading, Mass.: Addison-Wesley, 1993.

IN CONTEXT

This essay first appeared in the May–June 1991 issue of *American Scientist.* The article from *The Chronicle of Higher Education* that prompted the essay was called to my attention by Brian Hayes and Rosalind Reid.

Billington, David P. *The Innovators: The Engineering Pioneers Who Made America Modern.* New York: Wiley, 1996.

———. *The Tower and the Bridge: The New Art of Structural Engineering.* New York: Basic Books, 1983.

Cutcliffe, Stephen H., and Robert C. Post, eds. *In Context: History and the History of Technology: Essays in Honor of Melvin Kranzberg.* Bethlehem, Pa.: Lehigh University Press, 1989.

Florman, Samuel C. *The Civilized Engineer.* New York: St. Martin's Press, 1987.

———. *The Introspective Engineer.* New York: St. Martin's Press, 1996.

Hughes, Thomas P. *American Genesis: A Century of Invention and Technological Enthusiasm, 1870–1970.* New York: Viking, 1989.

Rosenberg, N., and W. G. Vincenti. *The Britannia Bridge: The Generation and Diffusion of Technological Knowledge.* Cambridge, Mass.: MIT Press, 1978.

Vincenti, Walter G. *What Engineers Know and How They Know It: Historical Studies in the Nature and Sources of Engineering Knowledge.* Baltimore: Johns Hopkins University Press, 1990.

MEN AND WOMEN OF PROGRESS

A good deal of the background information for this *American Scientist* column (May–June 1994) comes from anonymous documents in vertical files at the Smithsonian Institution and at Cooper Union, copies of which were kindly provided by their curators and librarians. I am grateful to Egle Zygas of the Smithsonian Institution's Cooper-Hewitt, the National Design Museum, for calling my attention to the museum's survey of patent models published in the catalog *American Enterprise.* Cooper Union was a most hospitable institution when I showed up unannounced one spring afternoon asking to view its *Men of Progress,* and Provost Edward Colker kindly interrupted his work to give me a private showing. I am also grateful to Harold R. Murdock, who after reading this essay in *American Scientist* informed me that Sartain's engraving was presented to subscribers of *Scientific American* and showed me his copy of that version of *Men of Progress.*

American Society of Mechanical Engineers. *Mechanical Engineers Born Prior to 1861: A Biographical Dictionary.* New York: American Society of Mechanical Engineers, 1990.

Commissioner of Patents, comp. *Women Inventors to Whom Patents Have Been Granted by the United States Government, 1790 to July 1, 1888.* Washington, D.C.: U.S. Government Printing Office, 1888.

Cooper-Hewitt Museum. *American Enterprise: Nineteenth-Century Patent Models.* New York: Cooper-Hewitt Museum, 1984.

Cutcliffe, Stephen H., and Robert C. Post. *In Context: History and the History of Technology. Essays in Honor of Melvin Kranzberg.* Bethlehem, Pa.: Lehigh University Press, 1989.

Hindle, Brooke, and Steven Lubar. *Engines of Change: The American Industrial Revolution, 1790–1860.* Washington, D.C.: Smithsonian Institution Press, 1986.

Macdonald, Anne L. *Feminine Ingenuity: Women and Invention in America.* New York: Ballantine Books, 1992.

[Skirving, John.] *Key to the Engraving of Men of Progress—American Inventors. Engraved on Steel by John Sartain after the Original Painting by C. Schussele, of Philadelphia.* Germantown, Pa.: John Skirving [ca. 1863].

Stanley, Autumn. *Mothers and Daughters of Invention: Notes for a Revised History of Technology.* Metuchen, N.J.: Scarecrow Press, 1993.

SOIL MECHANICS

This column first appeared in *American Scientist* for September–October 1996. I am grateful to Ralph Peck and to Richard E. Goodman, who is working on a biography of Terzaghi, for their helpful comments on the manuscript of the column.

Bjerrum, L., et al. *From Theory to Practice in Soil Mechanics: Selections from the Writings of Karl Terzaghi.* New York: Wiley, 1960.

Donnicliff, John, and Don U. Deere, eds. *Judgment in Geotechnical Engineering: The Professional Legacy of Ralph B. Peck.* Vancouver, B.C.: BiTech Publishers, 1991.

Jumikis, Alfreds R. *Introduction to Soil Mechanics.* Princeton, N.J.: Van Nostrand, 1967.

Leonoff, Cyril E. *A Dedicated Team: Klohn Leonoff Consulting Engineers, 1951–1991.* Vancouver, B.C.: Klohn Leonoff, 1994.

Proceedings of the International Conference on Soil Mechanics and Foundation Engineering. Cambridge, Mass.: Harvard University, 1936.

Terzaghi, Charles. "Old Earth-Pressure Theories and New Test Results." *Engineering News-Record,* September 30, 1920, pp. 632–37.

——. "Principles of Soil Mechanics" [in eight parts]. *Engineering News-Record,* November 5, 1925, etc., p. 742, etc.

———•———

Terzaghi, Karl. *Theoretical Soil Mechanics.* New York: Wiley, 1943.

Terzaghi, Karl, and Ralph B. Peck. *Soil Mechanics in Engineering Practice.* New York: Wiley, 1948.

IS TECHNOLOGY WIRED?

The core of this essay was prepared to accompany a portfolio of photographs from the collections of the Library of Congress showing radio broadcasting and listening in America in the 1920s. The shorter essay appeared in *Civilization,* the magazine of the Library of Congress, in February–March 1997. At least one reader took exception to the omission of the name of Nikola Tesla, whom the U.S. Supreme Court declared in 1943 to be the inventor of radio. My essay was not meant to refute this but to highlight Marconi's role in promoting the invention's potential.

In addition to the following references, manuscripts and course notes provided by Karl D. Stephan of the University of Massachusetts supplied helpful background information.

Billington, David P. *The Innovators: The Engineering Pioneers Who Made America Modern.* New York: Wiley, 1996.

Carswell, Charles. *The Building of the Delaware River Bridge, Connecting Philadelphia, Pa., and Camden, N.J.* Burlington, N.J.: Enterprise Publishing, 1926.

Douglas, Susan J. *Inventing American Broadcasting, 1899–1922.* Baltimore: Johns Hopkins University Press, 1987.

Mannes, George. "The Birth of Cable TV." *American Heritage of Invention & Technology,* Fall 1996, pp. 42–50.

Schiffer, Michael Brian. *The Portable Radio in American Life.* Tucson: University of Arizona Press, 1991.

HARNESSING STEAM

This essay first appeared in *American Scientist* for January–February 1996.

Burke, John G. "Bursting Boilers and Federal Power." *Technology and Culture* 7 (1966): 1–23.

Cross, Wilbur. *The Code: An Authorized History of the ASME Boiler and Pressure Vessel Code.* New York: American Society of Mechanical Engineers, 1990.

Rueth, Nancy. "Ethics and the Boiler Code." *Mechanical Engineering,* June 1975, pp. 34–36.

Scherer, F. M. "Invention and Innovation in the Watt-Boulton Steam-Engine Venture." *Technology and Culture* 6 (1965): 165–87.

Sinclair, Bruce. *A Centennial History of the American Society of Mechanical Engineers, 1880–1980.* Toronto: University of Toronto Press, 1980.

Turner, Roland, and Steven L. Goulden. *Great Engineers and Pioneers in Technology.* Volume I: *From Antiquity Through the Industrial Revolution.* New York: St. Martin's Press, 1981.

Vitruvius. *The Ten Books on Architecture,* trans. Morris Hicky Morgan. New York: Dover, 1960.

THE *GREAT EASTERN*

This essay, without much of the material about Scott Russell that is included here, first appeared in *American Scientist* for January–February 1992. The Scott Russell material appears in *American Scientist* for January–February 1998.

Anonymous. "Looking for Mr Brunel." *The Economist,* December 2, 1989, p. 70.

Beaver, Patrick. *The Big Ship.* London: Hugh Evelyn, 1969.

Beckett, Derrick. *Brunel's Britain.* Newton Abbot, Devon, Eng.: David & Charles, 1980.

Brunel, Isambard. *The Life of Isambard Kingdom Brunel, Civil Engineer.* London: Longmans, Green, 1870.

Buchanan, R. A. "The *Great Eastern* Controversy: A Comment." *Technology and Culture* 24 (1980): 98–106.

Emmerson, George S. "The *Great Eastern* Controversy: In Response to Dr. Buchanan." *Technology and Culture* 24 (1983): 107–13.

———. *John Scott Russell: A Great Victorian Engineer and Naval Architect.* London: John Murray, 1977.

———. "L. T. C. Rolt and the *Great Eastern* Affair of Brunel versus Scott Russell." *Technology and Culture* 21 (1980): 553–69.

Fairbairn, William. *Treatise on Iron Ship Building: Its History and Progress.* London: Longmans, Green, 1865.

Pugsley, Sir Alfred, ed. *The Works of Isambard Kingdom Brunel: An Engineering Appreciation.* London: Institution of Civil Engineers, 1976.

Rolt, L. T. C. *Isamard Kingdom Brunel.* Hammondsworth, Middlesex, Eng.: Penguin, 1957.

Russell, J. Scott. *The Modern System of Naval Architecture,* 3 vols. London: Day & Son, 1865.

DRIVEN BY ECONOMICS

This essay first appeared in *American Scientist* for November–December 1991.

Hutchings, David F. *QE2–A Ship for All Seasons.* Southampton: Kingfisher Railway Productions, 1988.

Miller, William H., Jr. *The Great Luxury Liners, 1927–1954: A Photographic Record.* New York: Dover, 1981.

Orlebar, Christopher. *The Concorde Story: Ten Years in Service.* Twickenham, Middlesex, Eng.: Temple Press, 1986.

Rolt, L. T. C. *Isambard Kingdom Brunel.* Harmondsworth, Middlesex, Eng.: Penguin, 1957.

Wohleber, Curt. "The Annihilation of Time and Space." *American Heritage of Invention & Technology,* Spring/Summer 1991, pp. 20–26.

THE PANAMA CANAL

This *American Scientist* column (January–February 1993) was written in May 1992 on board the S.S. *Costa Riviera*, during a voyage that included an Atlantic-to-Pacific transit of the Panama Canal.

Abbot, Henry L. *Problems of the Panama Canal: Including Climatology of the Isthmus, Physics and Hydraulics of the River Chagres, Cut at the Continental Divide, and Discussion of Plans for the Waterway, with History from 1890 to Date.* New York: Macmillan, 1907.

Keller, Ulrich. *The Building of the Panama Canal in Historic Photographs.* New York: Dover, 1983.

Mack, Gerstle. *The Land Divided: A History of the Panama Canal and Other Isthmian Canal Projects.* New York: Knopf, 1944.

McCullough, David. *The Path Between the Seas: The Creation of the Panama Canal, 1870–1914.* New York: Simon & Schuster, 1977.

THE FERRIS WHEEL

I am grateful to Norman Anderson for his critical reading of this column when it was in press at *American Scientist* for the May–June 1993 issue. His large and wonderful illustrated history of Ferris wheels, which came to my attention only at that time, is a wealth of fact and anecdote.

Anderson, Norman D. *Ferris Wheels: An Illustrated History.* Bowling Green, Ohio: Bowling Green State University Popular Press, 1992.

Anderson, Norman D., and Walter R. Brown. *Ferris Wheels.* New York: Pantheon, 1983.

Burg, David F. *Chicago's White City.* Lexington: University Press of Kentucky, 1976.

Funderburg, Anne. "America's Eiffel Tower," *American Heritage of Invention & Technology,* Fall 1993, pp. 8–14.

Jones, Lois Stodieck. *The Ferris Wheel.* Reno, Nev.: Grace Dangberg Foundation, 1984.

Snyder, Carl. "Engineer Ferris and His Wheel." *The Review of Reviews,* September 1893, pp. 269–76.

HOOVER DAM

This *American Scientist* column appeared in the November–December 1993 issue.

Boyle, Robert H. "Life–or Death–for the Salton Sea?" *Smithsonian,* June 1996, pp. 87–96.

Hoover, Herbert. *Memoirs: The Cabinet and the Presidency, 1920–1933.* New York: Macmillan, 1952.

Mead, Elwood. "Hoover Dam." *Civil Engineering,* October 1930, pp. 3–8.

Moeller, Beverley Bowen. *Phil Swing and Boulder Dam.* Berkeley: University of California Press, 1971.

Stevens, Joseph E. *Hoover Dam: An American Adventure.* Norman: University of Oklahoma Press, 1988.

U.S. Bureau of Reclamation. *Hoover Dam, Power Plant and Appurtenant Works: Specifications, Schedule, and Drawings.* Washington, D.C.: U.S. Department of the Interior, 1930.

THE CHANNEL TUNNEL

This *American Scientist* column appeared in the September–October 1994 issue.

Byrd, Ty. *The Making of the Channel Tunnel.* London: Thomas Telford, 1994.

Byrne, Robert. *The Tunnel.* New York: Harcourt Brace Jovanovich, 1977.

Davidson, Frank P. *Macro: A Clear Vision of How Science and Technology Will Shape Our Future.* New York: Morrow, 1983.

Finch, James Kip. *Engineering and Western Civilization.* New York: McGraw-Hill, 1951.

Sandström, Gösta E. *Tunnels.* New York: Holt, Rinehart, & Winston, 1963.

Sargent, John H. "Channel Tunnel Project." *Journal of Professional Issues in Engineering* 114 (1988): 376–93.

Whiteside, Thomas. *The Tunnel Under the Channel.* New York: Simon & Schuster, 1962.

THE PETRONAS TOWERS

This *American Scientist* column (July–August 1996) derived much information and insight from the work of Charles H. Thornton, who visited Duke University in February 1996 and gave a compelling lecture on the project. I am also grateful to Leonard M. Joseph, vice president of Thornton-Tomasetti Engineers, for his comments on a draft of the column, and to Peter H. Stauffer for his clarification of points relating to the Malaysian language and the geology of the Kenny Hill formation.

Gargan, Edward A. "A Boom in Malaysia Reaches for the Sky." *The New York Times,* February 2, 1996, pp. C1, C3.

Reina, Peter, et al. "Malaysia Cracks Height Ceiling as It Catapults into Future." *Engineering News-Record,* January 15, 1996, pp. 36–54.

Robison, Rita. "Malaysia's Twins." *Civil Engineering,* July 1994, pp. 63–65.

———. "The Twin Towers of Kuala Lumpur." *IEEE Spectrum,* October 1995, pp. 44–47.

Thornton, Charles H., et al. "High-Strength Concrete for High-Rise Towers." *Proceedings, American Society of Civil Engineers Structures Congress XIV,* Chicago, April 1996, pp. 15–18.

List of Illustrations

List of Illustrations

Index